This case study examines the interrelationship between mathematics and physics in the work of one of the major figures of the Scientific Revolution: the Dutch mathematician, physicist, and astronomer, Christiaan Huygens (1629–1695). Joella Yoder details the creative interaction that led Huygens to invent a pendulum clock that theoretically beat absolutely uniform time, to measure the constant of gravitational acceleration, to analyze centrifugal force, and to create the mathematical theory of evolutes.

In the second half of the book, Dr. Yoder places Huygens's work in the context of his time by examining his relationship with other scientists and the priority disputes that sometimes motivated his research. The role of evolutes in the history of mathematics is analyzed; the reception of Huygens's masterpiece, the *Horologium Oscillatorium* of 1673, is described; and finally, the part that Christiaan Huygens played in the rise of applied mathematics is addressed.

Unrolling time

UNROLLING TIME

*Christiaan Huygens and
the mathematization of nature*

Joella G. Yoder

The right of the
University of Cambridge
to print and sell
all manner of books
was granted by
Henry VIII in 1534.
The University has printed
and published continuously
since 1584.

Cambridge University Press

Cambridge

New York   New Rochelle   Melbourne   Sydney

Published by the Press Syndicate of the University of Cambridge
The Pitt Building, Trumpington Street, Cambridge CB2 1RP
32 East 57th Street, New York, NY 10022, USA
10 Stamford Road, Oakleigh, Melbourne 3166, Australia

First published 1988

Printed in the United States of America

Library of Congress Cataloging-in-Publication Data
Yoder, Joella Gerstmeyer, 1944–
Unrolling time : Christiaan Huygens and the mathematization of
nature / Joella G. Yoder.
p.   cm.
Bibliography: p.
ISBN 0 521 34140 X

1. Mathematics – Early works to 1800 – Collected works.   2. Science –
Mathematics – Early works to 1800 – Collected works.   I. Title.
QA3.Y63   1988
510′.92′4 – dc19   88-6975
CIP

British Library Cataloguing in Publication Data
Yoder, Joella G.
Unrolling time.
1. Physics.   Huygens, Christiaan, 1629–1695
I. Title
530′.092′4

ISBN 0 521 34140 X

*With great affection and appreciation
I dedicate this book to my mother, Vera W. Lane,
for whom the education of her daughter
was always first priority*

Contents

Preface

Any attempt to reconstruct the process by which a great mind discovered an important concept becomes itself a process of discovery, and the recounting of the reconstruction becomes a mystery tale twice told. The detritus of the subject's activities – the calculations in the margins, the remains of former values poking out from behind ink blotches – these are the clues left to the historical detective. Shall I tell you of the manuscript that had to exist, although there was no trace of it in the collected works, simply because the derivation it contained was the "logical" next step? Eureka, I found it – or rather he did.

It is incumbent upon me to explain the way that I present the evidence gathered during my pursuit of the elusive Chr. Huygens. For obvious reasons I have adhered as much as possible to his original derivations and avoided the finished propositions. I have rarely presented his arguments verbatim, however, and this raises the possibility that I have misinterpreted what is on the page. In particular, for the sake of rendering the arguments more accessible, I have made some concessions to readability. Modern notation for such entities as pi and the equal sign are introduced, except in one case where I wish to make a point regarding the notation. A potentially more serious substitution is my use of algebraic notation in geometric arguments. Huygens stated everything verbally when he was in his "geometric mode" and used symbols such as the radical sign only when he switched to his "algebraic mode." Facile mathematician that he was, he switched back and forth between the two modes as his needs changed within the same problem, a procedure that even the most devout purist would not duplicate. Hence, you will find the verbal arguments presented in an algebraic form never used by him; for example, "the subduplicate proportion of AB to CD" is rendered as $\sqrt{AB}/\sqrt{CD}$. What you will not find is an integral sign or an infinite sum sign or any other symbol that embodies concepts developed in more modern times. One word that does appear often and needs some justification is "infinitesimal," which I use as a shorthand expression for an arbitrarily small length. Huygens did use equivalents such as *particula* or, in the case of an infinite sum, *infinita considerata multitudine,* so that I do not feel that I am violating the spirit of his work or that of his contemporaries by using the word.

Fewer compromises have been made regarding the figures, only because the originals for each derivation have been included, even in cases where a redrawn version is provided for clarity's sake. However, note that, in order to maintain the configuration of the figures as drawn by Huygens and yet translate his mathematics into recognizable formulas in the notes, I have taken the vertical axis as the *x* axis in most cases.

Because the intent of this book is to show a master of the Scientific Revolution at work, as many manuscripts have been reproduced as cost and legibility have allowed. Since so much of the material is drawn from the working papers, including manuscripts not reproduced by the editors of the *Oeuvres complètes,* I have decided for consistency to provide my own translations of all passages from Huygens's work, even in those few cases where published translations exist, particularly including the recent English translation of the *Horologium Oscillatorium* done by Richard Blackwell. Finally, on the level of real minutia, all dates are given according to the Gregorian calendar already adopted on the Continent during Huygens's time, unless specifically noted as old style (o/s).

This book would not exist had it not been for the help of two venerable Dutch institutions. The University of Leiden has been the keeper of Huygens's manuscripts for nearly three centuries, and I thank Dr. P. F. J. Obbema and his excellent staff of the Department of Western Manuscripts for their assistance. All photographic reproductions of the manuscripts contained herein were provided courtesy of the library at Leiden. Equally essential to Huygens scholarship is the *Oeuvres complètes de Christiaan Huygens* (1888–1950) published by the Hollandsche Maatschappij der Wetenschappen (Holland Society of Sciences). Although I may sometimes disagree with the editors, for example, regarding a specific manuscript's date, I do not wish to denigrate their very substantial achievement.

I offer many thanks to the members of the society not only for their support of the original publication of the works but also for their continuing encouragement of research on Huygens. As I have already intimated, quotations are taken from the *Oeuvres* when possible; likewise reproduced from the collection are those figures belonging to the published treatises (Figs. 1.1, 8.1, and 8.4). I also thank the libraries and staffs of the University of Washington, the University of Wisconsin, Schaumburg (Illinois) Township, and Northwestern University for all the many services rendered over the course of my research.

This book has had many readers who encountered it in its various forms as it progressed from dissertation to printed treatise. Their discerning eyes and questioning minds righted egregious errors and prompted reassessments. Somehow a mere thank you seems meager recompense but will have to do. I gratefully acknowledge the help of Floris Cohen, Donald Crowe, James Evans, Mordechai Feingold, Thomas Hankins, David Lindberg, Daniel Siegel, and William Yoder. Additional thanks are owed to those who provided encouraging words from the sidelines – the members of the Seminar in History of Science at the University of Washington, as well as Rima Apple, Mary Robertson, and, once again, my supportive husband, Bill.

1

Introduction

In the era bounded by Galileo's *Dialogo* of 1632 and Newton's *Principia* of 1687, science changed. Observation, even when performed with enough care to be called experimentation, gave way to rigorous mathematical analysis as the primary approach to physical phenomena. Whereas Galileo aimed to instruct laymen about his view of the world order by means of plausible arguments and analogies, only an experienced mathematician could hope to understand the world picture envisioned by the *Principia*. This mathematization of physics was a defining element of that intellectual upheaval we call the Scientific Revolution, and the requirement, still imposed today, that a theoretical physicist be an able mathematician stems from a tradition that flowered in the seventeenth century.

Such sweeping change cannot be attributed to one particular moment or person. Yet the development of this interrelationship between mathematics and physics has remained too long in the realm of vague generalizations, whose validity has yet to be substantiated by a careful comparison with actual events. A new difficulty arises, however, because the particulars against which any generalization must be tested are not well documented. It is the latter deficiency that this book addresses by focusing on a specific person and event in the development of mathematical physics during the seventeenth century. This is a modest endeavor, designed not to explain the greater phenomenon but to provide a case study that any general account must encompass. The person is Christiaan Huygens; the event is his creation of the theory of evolutes.[1]

Preeminent mathematician, physicist, and astronomer, Christiaan Huygens (1629–95) was one of the major figures of the Scientific Revolution. Second son of the great Dutch poet and diplomat Constantijn Huygens, he was introduced at a very young age into a learned, cosmopolitan society and rapidly distinguished himself in mathematics and observational astronomy. His early achievements included studies on classical mathematics in the Archimedean and Apollonian traditions, an approximation of pi, the first printed treatise on probability, the discovery of one of Saturn's moons (Titan), the correct explanation of Saturn's varying profile (his ring hypothesis), and an unpublished treatise on the mechanics of impact. By his midthirties he had acquired such international acclaim that he was called to the court of Louis XIV to participate in the formation of the Académie Royale des Sciences, and there he remained for almost twenty years, except for trips home during periods of debilitating illness. Cited in modern histories of science primarily for his wave theory of light, his work on centrifugal force, his analysis of percussion, and the Huygens ocular for telescopes, he is regarded as the last great proponent of the mechanical philosophy (usually equated with Cartesianism).[2]

The design and development of clocks was one problem that interested Huygens throughout his life. In 1657, he created a clock whose advance was regulated by a pendulum, and consequently he has usually been designated the inventor of the pendulum clock.[3] Galileo Galilei had also considered using the pendulum as a timekeeper and had even given his son instructions on how to build a clock regulated by a swinging rod, a task never completed. Both Huygens and Galileo hoped that the new design would greatly improve the accuracy of astronomical measurements and make possible a precise determination of longitude at sea. Both scientists have partisans claiming priority of invention for their candidate's design, although still other enthusiasts argue that Leonardo da Vinci invented the pendulum clock.[4] As with most mechanical devices, priority of invention depends on whether the emphasis and value are placed on the basic design, on the construction of a physical model, or on the accuracy of the mechanism once constructed. Huygens's claim rests on the last criterion, for in addition to mounting the pendulum on the clock in a better manner, he instituted features that guaranteed far greater accuracy – for example, the endless chain, which made it possible to wind the clock without disturbing its progress. In 1658, Huygens published his *Horologium,* a description of his most recent design incorporating these advances, and thereby popularized the pendulum clock.[5]

A little more than a year later, Huygens began planning a second edition of this *Horologium* in order to incorporate a wealth of discoveries and consequent designs that he had developed in the meantime. Many years passed before this new work, one of the masterpieces of seventeenth-century scientific literature, was finally published in 1673 under the title *Horologium Oscillatorium* (The pendulum clock). Much more than a mere description of a clock, as the earlier work had been, it was in fact a treatise on the accelerated motion of a falling body, as exemplified by the bob of a pendulum clock.[6]

The book is divided into five parts, the first describing the mechanical features of a clock designed by Huygens. This clock (Fig. 1.1) included the endless chain, a lens-shaped bob that minimized air resistance, the *curseur* weight that allowed fine adjustment of the period of swing, and a pair of plates that were curved in the shape of an inverted cycloid and mounted on either side of the pendulum.

The second part of the *Horologium Oscillatorium* is a series of propositions on gravitational fall, beginning with free fall, including linear fall along inclined planes, and ending with fall along a curved path. The culminating proposition is Huygens's proof that a body falling along an inverted cycloid (Fig. 1.2) reaches the bottom in a fixed amount of time, irrespective of the point on the path at which it begins its fall. In other words, the cycloid is isochronous.[7]

The third section of Huygens's great work introduces his theory of evolutes, a mathematical correspondence between curves that, among other applications, allows one to find the length of a curve. Using evolutes, Huygens justifies his introduction of the curved plates to the clock that he describes in Part 1, for he proves mathematically that the cycloidal-shaped plates will force the bob of the pendulum to move along the isochronous cycloidal path. Thus, ideally, the pendulum will keep uniform time regardless of how wide it swings, as he has shown in Part 2.

The fourth, and longest, section of the *Horologium Oscillatorium* deals with the physical, rather than the ideal, pendulum. Here Huygens presents his theory of the compound pendulum, in which the motion of a pendulum with mass distributed along its length is compared with that of an ideal simple pendulum of weightless cord and point-mass bob.

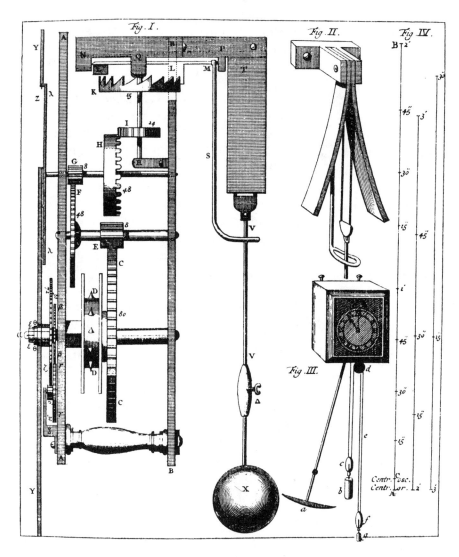

Figure 1.1. The clock of the *Horologium Oscillatorium*.

The last part of the book introduces a second timepiece, one that is a variant of a conical clock in which the pendulum, instead of swinging, rotates about a vertical axis. As with the cycloidal pendulum of Part 1, the bob is kept on an isochronous path by a curved plate whose shape is also determined by the theory of evolutes.

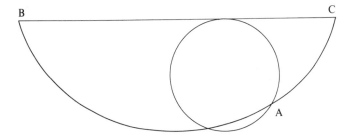

Figure 1.2. An inverted cycloid, drawn by tracing the path of a fixed point *A* on the generating circle's rim as it rolls counterclockwise along a straight line.

Following the description of this clock Huygens lists, without proofs, thirteen theorems on centrifugal force that form the theoretical justification of the pendulum's motion, in the same way that the theorems of Part 2 validate the clock of Part 1. The proofs of these theorems eventually appeared in the posthumously published *De Vi Centrifuga*.

The contents of the *Horologium Oscillatorium* are tied together much more closely than has ever been hinted in previous literature. Except for Part 4, which was not written until 1664, the entire treatise was essentially developed in a three-month period beginning in October 1659. During that time, Huygens proceeded rapidly, almost inexorably, from one creative event to another until the major theorems of the *Horologium Oscillatorium,* and *De Vi Centrifuga,* were revealed. This progression of ideas can be reconstructed in considerable detail by a careful examination of the massive evidence in the *Oeuvres complètes de Christiaan Huygens* and by a return to the original manuscripts, some of which are not included in the *Oeuvres complètes.*[8] A reconstruction of this period of great creativity is first and foremost a study of Huygens's method of research, and as such affords a detailed examination of the interaction between his mathematics and his physics.

Huygens's theory of evolutes, the key mathematical concept to emerge from this work of 1659, provides a natural focal point for such a study, since it was as much a physical as a mathematical idea, both in its roots and in its applications. This duality is evident in the definitions of the evolute and its companion, the involute. Although viewed today as a feature of pure mathematics, the evolute was originally conceived by Huygens in mechanical terms, and

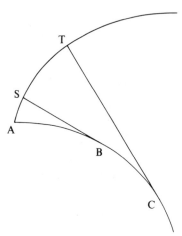

Figure 1.3. The mechanical evolution of an involute from the unrolling of its evolute.

its origin is reflected in its very name, which comes from the participle *evolutus* (unrolled) of the Latin verb *evolvere* (to unroll).[9] Given a curve *ABC* (Fig. 1.3), called the evolute, it is "unrolled" by the simple expedient of fitting a thread (*filum*) to its shape and carefully unwinding the thread from one end, with the freed end of the thread always pulled taut. The end of the thread, which begins at *A* and moves out toward *S* and thence to *T* as the thread unrolls, traces out a companion curve, which Huygens always referred to as "that drawn by the unrolling" (*descripta ex evolutione*) and which modern mathematics labels the involute.[10]

Although Huygens defined the relationship between evolute and involute mechanically, in practice he derived the evolute mathematically from a given involute. Given a curve *MPQ* (Fig. 1.4), where *Q* is infinitesimally close to *P*, the intersection *N* of the normal to *P* (the line perpendicular to the tangent at *P*) and the normal to *Q* is presumed to lie on the evolute. In modern terminology, never used by Huygens, the evolute is the locus of the instantaneous centers of rotation or curvature of the involute. The two approaches – mechanically deriving the involute and mathematically deriving the evolute – are equivalent, and one of Huygens's first tasks in Part 3 of the *Horologium Oscillatorium* is to show the uniqueness of the relationship between evolute and involute, by proving that the tangents to the evolute (the mechanical approach) are normals to the involute (the mathematical avenue).[11] Note that any curve can function as either an evolute or an involute; its role depends on the circumstances of the problem at hand.

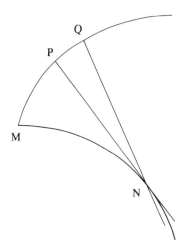

Figure 1.4. The mathematical derivation of
an evolute from the normals to its involute.

Evolutes grew out of Huygens's studies of the pendulum; the
complete theory was a product of his mathematical response to a
physical question. However, evolutes formed only one part of a
rich matrix of mathematical techniques that Huygens used to attack the problems presented him in late 1659. Those techniques and
how they helped him deal successfully with physical questions are
a major concern of this book.

A caveat is therefore in order: Mathematics ahead! I have tried to
avoid introducing extraneous mathematical material, and in particular I have minimized the introduction of modern equivalents to
the techniques used by Huygens, with most of the exceptions relegated to the notes as shorthand aids. This procedure should (1) ag-
gravate those mathematical adepts who seek an easy understanding
of his solutions, (2) please my fellow purists who feel that historical accuracy and insight are lost when results are couched in modern
terms (see the Preface for a confession regarding compromises),
and (3) bore those who cannot understand the mathematics in whatever format it is presented. For the sake of the last group, I have
tried to make sure that the derivations can be skipped or at least
skimmed without loss of the underlying story. However, the purpose
of this book is to demonstrate Huygens's mathematics at work in
the formation of his physics and vice versa, and thus the mathematics cannot be relegated to the notes. On the contrary, focusing on
his original derivations yields a rare glimpse of creativity in action.

The question that initiated Huygens's intensive period of research
at the end of 1659, which culminated in the theory of evolutes and

the cycloidal-pendulum clock, is easy to state, appearing almost inconsequential. How it leads to such fertile results reveals a master of seventeenth-century science in the process of discovery. The question: What is the constant of gravitational acceleration?

2

Accelerated motion: gravity

Christiaan Huygens's superior abilities in mathematics and mechanics were already evident in 1646, when he was but seventeen years old. Two examples of his scientific maturity are particularly significant for his future development. The first is a study of the catenary, which Huygens most likely undertook after having read Albert Girard's annotated edition of Simon Stevin's mathematical works, one of the books recommended to Huygens by his teacher as worthy of study.[1] In this work, Girard erroneously claims that the curve formed by a chain hung from both ends, called the catenary, is a parabola.[2] By viewing the hanging chain as a discrete set of equal weights distributed uniformly along a parabola, Huygens was able to disprove Girard's claim.[3]

At about the same time, Huygens undertook the study of a work by Juan Caramuel Lobcowitz, who claimed that the distances traversed by a falling body are proportional to the times elapsed.[4] Huygens refuted Lobcowitz's claim, showing by means of an arithmetic progression that, instead, the spaces are proportional to the squares of the times.[5] In a letter dated September 3, 1646, he boasts to his brother of his discovery, also claiming he can prove that, if the body is projected to one side instead of straight down, it will describe a parabola as it falls: "Of all this and infinitely more things that depend on it I have never known a demonstration before that of the discoverer, myself."[6] In a postscript the proud discoverer suggests that his brother show the letter to their father. Not long thereafter Christiaan obtained a copy of Galileo's *Discorsi e dimostrazioni matematiche intorno a duo nuove scienze*, in which the times-squared law and the parabolic shape of a projectile's path are presented, whereupon he canceled plans to write a small treatise on his discoveries: "I did not wish to write the Iliad after Homer had."[7]

In the meantime, however, his father had – as any proud father would – touted Christiaan's capabilities to his friend Marin Mersenne.[8] Well acquainted with the *Discorsi*, Mersenne was not at all sure that Galileo's treatment of fall was correct. Indeed, he begins his first letter to Christiaan by stating that he will not believe Huygens has the proof of falling bodies without seeing it himself. Moreover, he does not believe that a body goes through all degrees of speed when beginning its fall, as Galileo had claimed, concluding, "I add that the principles that Galileo has taken in all that he says about motion are scarcely firm."[9] But he was willing to be convinced, and to Christiaan's father he writes: "I anxiously await the demonstration of your son on the proportion of fall for heavy bodies, for perhaps he will have taken an approach independent of Galileo's suppositions."[10]

A few weeks later Christiaan sent Mersenne his derivations. In his analysis, he presumes that the ratio of the distance a body will fall from rest in a given amount of time to the distance fallen in the next equal unit of time will be a constant, irrespective of the unit of time chosen. From this implicit definition of "constant acceleration" he shows that, of all arithmetic or geometric progressions, only one can represent the increase in distance fallen, namely the arithmetic sequence $1, 3, 5, 7, 9, \ldots$, and thus the total distance fallen is as $1, 4, 9, 16, \ldots$.[11] The parabolic path of a projectile, he continues, follows easily from the known properties of the parabola and the fact that the total distance of fall is proportional to the square of the time. Huygens concludes by offering to send Mersenne his proof that the catenary is not parabolic in shape, a promise fulfilled a month later.[12] Mersenne's reply is enthusiastic: "I assure you that I have so greatly admired the finesse of your demonstration concerning falling bodies, that I believe that Galileo would have been delighted to have you as a defender of his opinion."[13]

These early episodes reveal that, by the time he began his correspondence with Mersenne, the young Huygens had already developed certain fundamental precepts and had successfully applied them to problems concerning the influence of gravity. That he accepted many of the principles associated with the mechanical philosophy he readily admitted to Mersenne.[14] His universe consisted of particulate matter in motion, rectilinear motion unless constrained or impacted. Such a metaphysics permits an easy translation of the physical into the geometric, substituting point for particle and line for trajectory or magnitude of some other physical quantity dependent on time. Cartesian physics is proof that acceptance of the mechanical philosophy does not necessarily lead to a mathematization of nature; the medieval latitude of forms is evidence that this mathematical association need not be accompanied by a profound mechanics. However, for Huygens the universe was both mechanical and mathematical.

Essential to his development as a mathematical physicist was his acceptance of the idealization of experience as taught by Galileo in the *Dialogo*. In criticizing Lobcowitz, Huygens notes:

Still, he has not considered motion abstractly enough, since he considers the motion of a stone or metal sphere falling through the air from on high; truly, far greater does the thought of Galileo agree with experience, excepting that the resistance of air impedes it. Thus, we will consider accelerated motion in a better way.[15]

With this comment, Huygens stated not only his immediate goal but also what was to be, in fact, a major research objective throughout his life – to explain accelerated motion in a better way than had been done previously.

The contact with Mersenne and the receipt of Galileo's *Discorsi* could only have reinforced his ambition. Galileo had presented the most complete account of fall to date, but he had not always given a convincing proof and had left many unresolved problems, as Mersenne was to point out by his continuing questions. Thus, Huygens's research took on a decidedly Galilean flavor. Even his study of percussion in the early 1650s, although primarily an analysis (and rejection) of Cartesian rules of impact, drew justification from Galileo's promise to present a treatise on the subject. Loosely transcribing the passage in the *Discorsi* where Galileo makes this pledge, Huygens adds that since Galileo never fulfilled his promise *he* will present his own treatise on percussion.[16] As will be seen, the *Horologium Oscillatorium* is the great culmination of Huygens's program of studying motion "in a better way," guided by the precepts of Galileo and the questions of Mersenne.

MERSENNE'S PROBLEM

Many of the topics Mersenne raised for discussion were not immediately pursued by the young Huygens but were taken up years after the death of his correspondent, whom, by the way, he never met. For example, in his first letter Mersenne had included a copy of Evangelista Torricelli's treatise on the cycloid, which captured Huygens's interest only in 1658, when, prodded by the challenges of Blaise Pascal, he undertook a study of the curve.[17] Likewise, Mersenne was unsuccessful in drawing his attention to the question of the center of oscillation of a vibrating body, a problem that was not attractive to Huygens until after he had developed his own interest in the motion of the nonideal pendulum. Huygens acknowledges his debt to Mersenne on this point in his introduction to the fourth part of the *Horologium Oscillatorium,* in which he masterfully deals with the problem of the compound pendulum.[18]

In 1659, Huygens took up another question previously handled by Mersenne, that of determining an accurate value for the constant of gravitational acceleration. In the seventeenth century, an era in which physical relationships were not yet expressed in formulas like $s = \frac{1}{2}gt^2$, this meant finding the distance traversed by a body in its first second of free fall, a value numerically equal to one-half the modern constant. The problem was compounded by the fact that an accurate means of measuring the second was not yet available, and thus each researcher first had to establish his timekeeper. (Ultimately the second is defined astronomically; $24 \cdot 60 \cdot 60$ seconds must make one day as measured by a star transit.)

Mersenne had been plagued by difficulties when he attempted to verify Galileo's claims regarding free fall. He wanted to find the length of a pendulum that would complete its swing (from one side to the other, but not back again) in 1 second and, then, to use that timepiece to determine the distance traveled in 1 second by a freely falling body. In his early experimental work Mersenne had concluded that this seconds-pendulum had a length of 3½ Parisian Royal feet, but later experiments led him to shorten this length to 3 feet. He was not completely satisfied with the new value, however, pointing out that the vibrations vary because the pendulum is not truly isochronous and is subject to the resistance of the air. In fact, when he used a pendulum of either length to time the fall of a ball, he got the same result: 12 feet of fall in one swing, 48 feet in two![19]

In his *Reflexiones Physico-mathematicae* (1647), Mersenne summarizes his difficulties and attempts to rectify the situation. He redoes his experiments, this time fixing a 3-foot pendulum to the wall

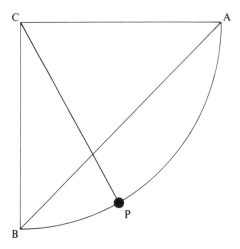

Figure 2.1. Mersenne's problem.

and holding the bob and a loose ball in the same hand so that they can be released simultaneously. By adjusting the height at which the bob and ball are released until the sound of the bob hitting the wall and the sound of the ball hitting the floor are heard simultaneously, he hopes to determine the distance the ball falls in the ½ second it takes the pendulum to reach the wall. Unfortunately, even with the improved technique, he reports achieving simultaneity of sound not only for a free fall of 3 feet (equivalent, by the times-squared law, to his earlier value of 12 feet in a full second) but for fall through any distance between 3 and 6 feet.[20]

The value he favors, 3 feet of free fall in ½ second, is the most vexing since it implies that the time of fall along the quarter arc of the circle swept out by the 3-foot pendulum is equal to the time of free fall through the radius (Fig. 2.1). But, Mersenne reasons, in the time that the bob falls the length of the quarter arc it should be able to fall perpendicularly a distance at least equal to the arc length, since fall along the perpendicular should always be as fast as, if not indeed faster than, descent along any other equally long path. Thus, he calculates, a seconds-pendulum of 3-foot length should actually correspond to a free fall of at least 5 feet in ½ second.[21] Indeed, when he then performs further experiments with this theoretical lower bound on the height of release, he does obtain a free fall of 5 feet 10 inches in ½ second.

However, a second theoretical attempt to analyze the problem leads him to doubt this result. Invoking Galileo's work on fall along chords of a circle, Mersenne notes that the time of fall along the quarter arc *AB* must be less than the time of fall along the chord *AB* spanning the arc. Further, Galileo's results on fall through inclined planes of equal vertical heights state that the time of fall along the chord *AB* is to the time of free fall along the radius *CB* as the diagonal of a square is to its side. Thus, the time of fall along the quarter arc should be less than $\sqrt{2}$ times the time of free fall through the radius.[22] Although a free fall of 5 feet 10 inches in ½ second does not violate this upper bound, Mersenne seems to feel that he has not convincingly analyzed the matter, for he abandons this value in favor of his original result of 3 feet of fall in the half-second swing of a 3-foot pendulum, despite its attendant paradox. Perhaps, he suggests, he has the wrong length for the seconds-pendulum just as he had previously erred in thinking it was 3½ feet. Perhaps it should be shortened to the 22 inches that his first analysis would claim is commensurate with a free fall of 3 feet in ½ second.[23] However, Mersenne did not redo the experiment with the shorter pendulum; rather, he laid the problem aside, apparently defeated.[24]

RICCIOLI'S ATTEMPT

At about the same time as Mersenne, Giambattista Riccioli also set out to measure with the aid of a pendulum the distances fallen by a body under the influence of gravity in various lengths of time. Among other questions, Riccioli wanted to test Galileo's claim that for falling bodies the distances increase in the same ratio as the odd numbers. Like Mersenne, he had difficulty arriving at a precise length for his seconds-pendulum. In fact, the pendulum with which he measured free fall was very short and swift, being $1^{25}/_{166}$ inches long and completing a swing in ⅙ second. With the aid of his Jesuit brothers, who among other things threw balls off the Torre degli Asinelli at various heights while chanting in synchrony with the pendulum in order to count its swings, he was able to confirm Galileo's claim. He summarizes the results of three separate trials in a table printed in Book 9 of his *Almagestum Novum*.[25] In fact, when the necessary conversions are performed, the values in the table lead to differing "constants" of acceleration, but he does explicitly record that when the pendulum beats 6 vibrations, that is, 1 second, a ball of clay will fall through the air a distance of 15 feet.[26]

In a now classic article entitled "An Experiment in Measurement," Alexandre Koyré discusses the experiments done by Mersenne and Riccioli to measure the gravitational constant. Koyré's thesis is that Huygens was able to arrive at the true value of the constant of gravitational acceleration by abandoning this experimental approach in favor of the theoretical, whereby Huygens achieved the exact formula for the gravitational constant in terms of the time of swing of a pendulum. Koyré concludes:

> We have seen Galileo, Mersenne, Riccioli endeavoring to construct a timekeeper in order to be able to make an experimental measure of the speed of the fall. We have seen Huygens succeed, where his predecessors had failed, and, by his very success, dispense with making the actual measurement. This, because his timekeeper is, so to say, a measurement in itself; the determination of its exact period is already a much more precise and refined experiment than all those that Mersenne and Riccioli have ever thought of.[27]

Although overstating the case – Huygens still had to make an actual measurement – Koyré is correct when he claims that Huygens found the constant of gravitational acceleration by developing new clocks that contained within their theoretical foundations a precise formula relating the constant to the period of the timekeeper. However, although essentially true, Koyré's remarks stem more from his own philosophical biases than from an examination of the relevant textual material. A closer study of Huygens's approach to the problem of measuring acceleration not only confirms Koyré's thesis but also reveals a fascinating pattern of discovery, a tightly knit sequence of questions and results. Indeed, after two months of concentrated activity beginning in late October 1659, Huygens not only had found the true value for the constant of gravitational acceleration and invented clocks by which to measure it, but had formulated a mathematical explanation of circular and gravitational motion far surpassing any work done by his predecessors.

3

Accelerated motion: centrifugal force

On October 21, 1659, twelve years after Mersenne had described his efforts to determine the constant of gravitational acceleration and the problems that he encountered, Christiaan Huygens repeated Mersenne's experiment. He accepted Mersenne's length of 3 feet for the seconds-pendulum but added an extra inch, thus compensating for the fact that his Rhenish foot was shorter than the Parisian Royal foot used by Mersenne, and found that a lead ball fell 3½ feet during the ½-second swing of the pendulum against the wall. Thus, by Galileo's times-squared law, it would fall 14 feet in 1 second, a value significantly different from Mersenne's 12 feet, even when the latter was converted to Rhenish feet (12 feet 5 inches).[1] Since Huygens also knew of Riccioli's value of 15 feet in 1 second, the discrepancy between the experimental values determined by himself, Riccioli, and Mersenne was readily noticeable. It is not surprising, therefore, that Huygens sought a different means of determining the constant of gravitational acceleration.

That same October day on which he tried Mersenne's experiment, Huygens undertook a study of centrifugal force, recording a list of propositions that a few weeks later he would develop into a more complete treatise, the work now called *De Vi Centrifuga*.[2] He was obviously still hunting for an accurate value for the constant of gravitational acceleration, for among the opening notes of the first draft is a reference to Riccioli's trials showing that gravity increases according to the odd numbers, followed by the phrases "Weight is the *conatus* [effort, tendency] to descend" and "concerning an accurate measure by means of the oscillations of a clock."[3]

The clock that Huygens would develop in order to measure the constant of gravitational acceleration was theoretically based on the close correspondence he envisioned between centrifugal force and gravity. In a separate passage, composed about the same time, Huygens writes:

The weight of a body is the same as the *conatus* of matter, equal to it and moved very swiftly, to recede from a center. He who holds it suspended prevents that matter from receding; he who allows it to fall offers to the same matter the capacity of receding from the center along the radius. Since however at the beginning it recedes from the center according to the odd numbers starting with one, inevitably it compels the heavy body to approach the center with a similarly accelerated motion, so that at the beginning these motions – the recession of matter from the center, and the approach of the falling body toward the center – are necessarily equal. And therefore, having found how much the body descends in a given time, for example, if it falls $\frac{3}{5}$ of a line [$\frac{1}{12}$ inch] in $\frac{1}{60}$ second, we will also know how much that matter ascends from the center, which certainly will also be $\frac{3}{5}$ of a line in $\frac{1}{60}$ second.

Hence the speed of matter is given by the radius of the earth.

Hence follows the centrifugal force in a smaller circle. Now, it ought to be seen how it resides in a thing and what determines the magnitude of this *conatus*. Certainly, the amount of recession in a fixed time depends on both the speed of gyration and the size of the circular path.[4]

Vaguely worded, the passage has usually been interpreted as an example of Huygens's early commitment to a Cartesian research program, particularly to a vortex explanation of gravity, in which the receding matter causes (compels) a downward (toward the center) motion of a body of equal size.[5] In fact, perhaps reflecting his own confusion about the actual mechanics of the swirling vortex, Huygens's description of this causal process is rather circular (no pun intended). The *conatus* of receding matter compels fall, but fall gives that matter the capacity to recede in the first place.[6] Whatever the causal mechanism between them, however, clearly in Huygens's view centrifugal force and gravity are reciprocal forces.

Written on a loose sheet and not part of his journal, this passage presumably predates Huygens's repetition of Mersenne's experiment, because the value cited corresponds to Riccioli's constant of gravitational acceleration rather than the value obtained by Huygens on October 21. The event that stimulated Huygens's interest in the problem of gravity and the consequent composition of this passage is purely a matter of speculation, since the sheet is undated.[7] However, the passage is almost a prospectus for the opening propositions of the original draft of *De Vi Centrifuga*, begun October 21, in which Huygens calculates changes in centrifugal force with respect to the radius of the circle and the speed of rotation. Thus, it is reasonable to assume that the passage was composed very close to October 21 and that it belongs to one continuous period of research.

And yet a very important change occurred in that narrow gap thus left between this preliminary passage and the beginnings of *De Vi Centrifuga.*[8]

An extensive discussion of the similarity of gravity and centrifugal force found at the beginning of the revised draft of *De Vi Centrifuga* reflects the change, while still maintaining the prospectus outlined in the earlier passage:

> Furthermore, whenever there are two bodies of equal weight each one held by a cord, if they have the same *conatus* due to accelerated motion, whereby they would pass through equal spaces in the same time, receding along the extension of the cord: we assert that the same tension is felt on their cords be they drawn downward or upward or in whatever direction.... And it ought to be measured by the initial motion, taking an arbitrarily small span of time.... Now let us see what and how much *conatus* there is in bodies bound to a cord or to a rotating circle as they recede from the center.[9]

He goes on to argue, in a very Galilean style, that the distances separating a body moving along its inertial path and the point on the circle from which it was released are to the times elapsed as the squares $1, 4, 9, 16, \ldots$. However, this is exactly the same relationship that holds between the times elapsed and the distances spanned by a freely falling body. Since the effects are the same, both centrifugal force and weight must engender the same kind of accelerated motion.

Gone is the Cartesian vortex (if, indeed, it existed in the earlier passage), and in its place gravity and centrifugal force are literally tied together by their similar effect on a resisting cord. A weight suspended by a cord exerts a tension on that cord; likewise, a body held in circular motion by a cord creates a tension on its restraint. In each case, the cord counteracts the body's natural tendency to move with an accelerated motion commensurate with the times-squared law, and the tension becomes a measure of the tendency to move. The dynamical problem is reduced to a "static" mode, in which Huygens measures the forces under question by means of the counterbalancing tensions. Finally and most significantly, because the "same tension is felt," be it from gravity or centrifugal force, the measure of one can become the measure of the other. Gravity and centrifugal force become reciprocal tensions.

The entire *De Vi Centrifuga* is phrased in terms of tensions, culminating in an investigation of the conical pendulum, which is nothing other than a body restrained by a cord and acted upon by both gravity and centrifugal force. Perhaps his desire, stated at the start of the treatise, to measure gravity more accurately by means of a pendulum clock prompted Huygens to view the problem in terms of tensions on the pendulum's cord. Yet much more probably the inspiration came from his repetition of Mersenne's experiment, which took place between his composition of the early passage and the first draft of *De Vi Centrifuga,* for Mersenne's embodiment of gravity in the pendulum naturally focuses attention on the cord's action in resisting the free fall of the bob.[10]

Excellent proof that Huygens did view Mersenne's experiment as a problem of tensions is provided by the penultimate proposition of the first draft of *De Vi Centrifuga,* in which Huygens translates Mersenne's comparison of free fall and pendular fall into a "statics" problem that compares the tension on a cord exerted by a hanging weight and the tension exerted by the weight if it were first to fall along the quarter arc of a circle.[11] Indeed, the first draft of *De Vi Centrifuga* can be interpreted as an attempt by Huygens to solve Mersenne's problem in its mathematical rather than its experimental guise. It was only the first of many solutions, for once Huygens performed the experiment he remained fixated on its ramifications. Thus, although it might not have initiated Huygens's study of gravity in 1659 and although it was too crude to yield a conclusive value for the constant of gravitational acceleration, Mersenne's experiment was crucial, because it inspired an approach to the immediate problem of measuring gravity that would lead to its solution and because it established a research topic that would lead Huygens beyond the immediate goal.

DE VI CENTRIFUGA

The correspondence between centrifugal force and gravity can be quantified, and the proposition that opens the first draft of *De Vi Centrifuga* states Huygens's basic formula. Although the derivation by which he arrived at the formula is not recorded, later propositions in both the first draft and the revision of *De Vi Centrifuga* provide the material for a reconstruction based on the proposition's accompanying diagram. He achieves the quantification by mathematically superimposing a parabola, which represents the path of a falling body that has been projected horizontally, onto a circle, around which a body is moving in a gravity-free situation. From his work for *De Circuli Magnitudine Inventa* (1654), Huygens

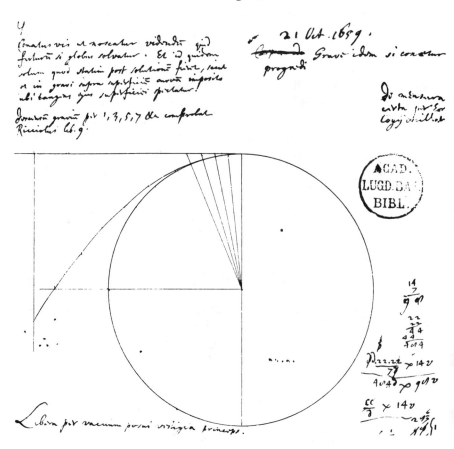

Figure 3.1. The beginning of *De Vi Centrifuga.*

knew that he could approximate any circle with a companion parabola whose *latus rectum* equals the diameter of the circle.[12] The diagram (Fig. 3.1) that accompanies the first proposition of *De Vi Centrifuga* pictures such a circle and its approximating parabola, along with the common tangent line to the one point at which they precisely agree and about which, within an infinitesimal distance on either side, the parabola can be considered to lie along the circle.[13]

To facilitate a reconstruction of Huygens's derivation of the formula that follows from this diagram, the figure must be labeled and an auxiliary line *CF* that is perpendicular to the tangent and describes the distance of the parabola from the tangent must be

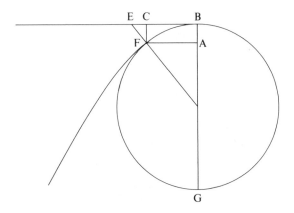

Figure 3.2. Huygens's basic formula: $EB^2 = EF \cdot BG$.

introduced (Fig. 3.2). By construction, the parabola has a *latus rectum* equal to the diameter BG of the circle, and therefore it is defined by the equation $CB^2 = CF \cdot BG$, where F, when close to B, is considered to lie not only on the parabola but also on the circle. In the time that a body moving uniformly around the circle BFG takes to reach F from B, that body, if released at B, would move inertially along the tangent line with a uniform rate to a point just shy of E, where E is the intersection of the extended radius through F and the tangent at B. Therefore, EF can be taken as the measure of the separation of the two paths and thus can be considered, relative to F's position, to be the distance through which the body moves away from F with an accelerated motion due to the centrifugal force in that same unit of time. Now, because F is considered to be very (infinitesimally) close to the point of release B, EF can be taken equal to CF and EB equal to CB. The definition of the given parabola, $CB^2 = CF \cdot BG$, and the identification of infinitesimal quantities yields $EB^2 = EF \cdot BG$.[14]

By considering that first infinitesimal moment that determines the motion of the released body and replacing the circle with its approximating parabola, Huygens achieved a quantitative expression for a body's separation from the circle and thus for its centrifugal force. In the time that it takes a body to move with uniform motion through an arc of the circle FB, which can be considered equal to EB when infinitely small, the body, if released at B, would move with an accelerated motion due to centrifugal force (as viewed from the moving point of release) through a distance *EF* equal to the square of the arc length *FB*^{2} divided by the diameter of the circle *BG*.

However, the approximating parabola is more than a mere mathematical equivalent to the circle, conveniently introduced to reduce a relationship involving centrifugal force to an easily stated equation; the parabola also represents the distance traveled by a freely falling body. In the same amount of time that the body takes to reach *E*, moved by centrifugal force, the body can fall through the distance *CF*. Again, for points very close to *B*, *CF* can be taken as equal to *EF*, and the motions induced by the two forces – the outward centrifugal force and the downward gravity – exactly counteract each other. Once the two forces, or more precisely the two tensions (these are virtual displacements), are shown to be equivalent, the formula can be extended. Specifically, in the time the body requires to move around the circumference πd of a circle of diameter *d* with a uniform motion, it can move through the distance $\pi^2 d$ $[=(\pi d)^2/d]$ with an accelerated motion due to centrifugal force, or, equivalently, it can fall under the influence of gravity through the same distance $\pi^2 d$.

The opening proposition of the preliminary *De Vi Centrifuga* summarizes this argument. It states that, if a body is constrained by a cord to move uniformly in a circular path of diameter *d* on a horizontal (gravity-free) plane, the tension on that cord due to the centrifugal tendency will equal the tension exerted by the weight of the body (were it suspended freely) if the body travels the circumference of the circle in the same time that it would fall through the distance $\pi^2 d$.[15] It follows, Huygens notes, that to move around a circle in twice the time and yet maintain the same tension requires a circle with a diameter four times as large.[16]

Using the constant of gravitational acceleration (14 feet of fall in the initial second) just obtained from his repetition of Mersenne's experiment, Huygens calculates the radius of the circle whose circumference a body would circuit in 1 second, with its centrifugal force equal to its weight; the value is 8½ Rhenish inches. Repeating the computation for a 24-hour period and comparing his answer with Snell's value for the radius of the earth, he finds that the radius of the earth would have to be hundreds of times larger than it is for the centrifugal force to equal the weight of a body, thereby allowing the body to fly off.[17]

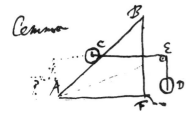

Figure 3.3. The mechanical principle:
wt. D/wt. $C = BF/AF$.

Two days later, on October 23, Huygens attempted Mersenne's experiment a second time, performing it with more attention to detail than before. He still accepted Mersenne's value for the length of the seconds-pendulum, but increased proportionally the length of his experimental pendulum to 6 feet 11 inches and thus slowed the rate of a full swing to 1½ seconds, thereby allowing ¾ second for the pendulum to hit the wall. Perhaps he felt that, in the larger span of time, he could more readily differentiate the striking of the two balls. His report concludes that a body falls 13 feet 8 inches in the first second, a smaller value than that obtained in his first trial but still a full foot longer than Mersenne's.[18] He adjusts his previous calculations in the draft of *De Vi Centrifuga* to reflect this new constant of gravitational acceleration, then continues his list of propositions on a body moving circularly under the restraint of a cord. The tension (*attractio*) on the cord is doubled as the length is doubled, the time being kept constant; the tension is quadrupled as the speed is doubled, the radius being kept constant; the tension (*attractionis vis*) is inversely proportional to the radius, the velocity being kept constant.[19] In modern notation, $F = mv^2/r$.

These first few propositions deal with circular motion in a horizontal plane and with the resulting tension on a cord that prevents the body from leaving a circular path. With his next proposition, Huygens moves from a planar to a spatial situation in which the weight of the circulating body is now considered along with its centrifugal force. This added physical dimension will allow Huygens to develop the mathematical theory necessary for the conical pendulum.

The proposition is a statement of the mechanical principle that the amount of horizontally acting force – viewed by Huygens as a hanging weight D pulling laterally by means of a pulley (Fig. 3.3) – necessary to sustain a body on an inclined plane is in the same ratio to the weight of the body C as the height of the plane to its base.[20] In particular, if the height equals the base, the weight on the

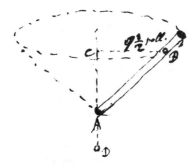

Figure 3.4. A ball in a rotating tube.

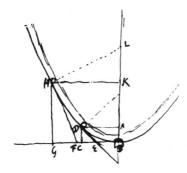

Figure 3.5. A ball in a rotating parabolic tube.

plane and the sustaining force are equal. Now, Huygens has already shown how to adjust the diameter of a circle to make the centrifugal force of a body equal its weight. In the propositions that follow he uses a variety of mechanisms to adjust both the inclined plane and the diameter of the circle in order that the centrifugal force D of a body revolving in a fixed circular path exactly counterbalance the downward pull of the weight C, thereby sustaining the body in its path.

The first mechanism is a simple combination of the circle and the plane (Fig. 3.4). According to Huygens's revised calculations, a ball moving on a circle of radius $8\frac{3}{10}$ inches in 1 second will have its centrifugal force equal to its weight.[21] If that ball is in a tube inclined at 45°, making one revolution per second, the centrifugal force will be enough to sustain the ball at an altitude of $8\frac{3}{10}$ inches (where the height equals the base), and thus the ball will not roll down the tube.

Likewise, if the tube is parabolic (Fig. 3.5), with a *latus rectum* of $16\frac{6}{10}$ inches, the ball will be held at an altitude A equal to one-

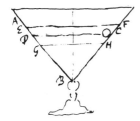

Figure 3.6. A ball in a rotating conical chalice.

fourth the *latus rectum*, because at that position the tangent to the parabola is inclined to 45° and the radius of the circle at that height is 8³⁄₁₀ inches. Huygens then notices that, in fact, at whatever altitude it is placed in the parabola the ball will not fall! At height K the ball H will have a centrifugal force larger than that at A by a factor of HK/DA, since centrifugal force is proportional to the radius of the circle being traveled if the time of revolution is kept constant. In addition, however, the force necessary to sustain the ball at H on the tangent plane HF is to its total weight as HG/GF, according to the mechanical principle already stated. But in a parabola HG/GF equals HK/DA, and therefore the two factors are equal, and the centrifugal force at H is precisely that necessary to sustain the weight of the ball at that altitude in the parabolic tube. Thus, the ball stays at whatever level it is placed, assuming the tube is rotating at a rate of one revolution per second. In general, Huygens goes on to note, any parabola will suffice as long as the rate of rotation is changed proportionally so that the broader the parabola the slower its rate of spin.[22]

Turning from spinning tubes, Huygens applies the same ideas to chalices, inside which the ball is kept in motion. Considering a conically shaped chalice (Fig. 3.6), he cites experiments (*experientia*) that show that, if a glass of this shape is turned swiftly, a ball will descend to a fixed height in the glass (determined by the speed of spin) and remain there, neither increasing nor decreasing its height. Moreover, in a paraboloidal chalice or paraboloidal mirror (Fig. 3.7), given the correct speed of spin (determined by the *latus rectum*), a ball will stay at whatever altitude it is released, for just as with the parabolic tube the centrifugal force is in equilibrium with the effective weight of the ball at any height. Thus, the ball will transit all circles on the spinning paraboloidal chalice in equal times, since to move faster (or slower) at one altitude would mean that

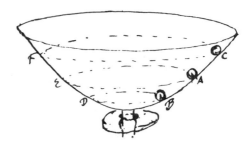

Figure 3.7. A ball completing all circles on a rotating parabolic chalice in the same time.

the centrifugal force is greater (smaller) than the effective weight at that height, and the ball would increase (decrease) its altitude until it was in equilibrium.[23]

Huygens had set out to find the correct value for the constant of gravitational acceleration by studying uniform circular motion. Very methodically approaching the subject, he had first considered centrifugal force in a planar situation in which the weight of the body was ignored. Then adding the extra dimension of weight, he had viewed the moving body as sitting on an inclined plane, kept there by a centrifugal force that counteracted the pull of gravity. The inclined plane then became a tangent to a paraboloid in which a ball completes every revolution in the same amount of time, no matter what its altitude. In this step-by-step fashion Huygens had discovered, without explicitly searching for it, the isochronism of the paraboloid of revolution. "If its rotations were counted, in this way an exact measure of time would be had, by means of a more accurate pendulum."[24]

THE CONICAL PENDULUM

The pendulum that the propositions in *De Vi Centrifuga* go on to describe, however, is one based not on the paraboloidal but on the simple conical model. A bob suspended from a cord at a fixed angle is moved in a circular path, so that its cord sweeps out the surface of a cone *ABE* (Fig. 3.8). (Note that if the angle were not fixed the bob would move on the surface of a sphere, not the isochronous

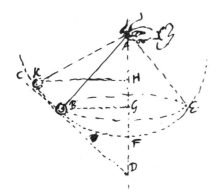

Figure 3.8. The conical pendulum.

paraboloid.) Huygens calculates that the length of the cord necessary to make the bob transit the circle in 1 second is $11\frac{8}{11}$ inches ($8\frac{3}{10}$ inches times $\sqrt{2}$), assuming the cord is kept at a 45° angle (angle *BAG*). At any other angle of inclination and thus altitude *H* for the same cord, he calculates, the bob will remain at altitude *H* if it travels the circle with a time equal to $\sqrt{HA}/\sqrt{GA}$, where *GA* equals *BG*, the radius of the seconds-circle, namely, $8\frac{3}{10}$ inches. In other words, the time of revolution of a conical pendulum is proportional to the square root of the height of the bob, as measured by its vertical distance *HA* from the vertex.[25] In particular, therefore, if the bob can be made to maintain the same vertical height, it will transit all circles at that altitude in the same time, irrespective of the angle the cord makes.

How much tension is felt by the cord of a conical pendulum? The bob can be visualized as sitting on an inclined plane *BE* tilted perpendicularly to the extended cord of the pendulum *AB* (Fig. 3.9, top). The centrifugal force thrusting out along the horizontal line *DBG* must counteract the tendency of the bob to move down this tangent plane because of its weight, and thus applying the mechanical principle he has previously invoked, Huygens argues that the centrifugal force must be to the weight of the bob as *BF* to *FE*. The combined pull on the cord can be viewed as two weights hanging at the point *B* (Fig. 3.9, bottom), one (the force of gravity) pulling downward and equal to the weight *H* of the bob and the other (the centrifugal force) pulling horizontally over a pulley and equal to *BF/FE* times the weight *H*. If the cord of the pendulum

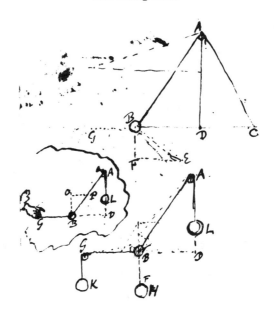

Figure 3.9. Tensions in the conical pendulum.

is now extended and passed over a pulley situated at the vertex *A*, the tension can be represented by a weight *L* that hangs from the end of the cord and counteracts the combined pull on *B*. Drawing on the mechanics of Simon Stevin, which obviously inspired his whole approach, Huygens concludes that the weight *L* is to the weight *H* as *AB* is to *AD*, which implies that the tension (*attractio*) is independent of the length of the cord *AB* and of the centrifugal force (weight *K*) and dependent only on the weight of the bob and the angle the cord makes with the perpendicular.[26]

These many geometric properties hold the information necessary to design a conical clock that maintains uniform time. First, the rate of rotation will be uniform if the bob can somehow stay at a constant vertical height. In a preliminary drawing (Fig. 3.10), Huygens sketches this fact and its consequence, namely that the excess cord of the pendulum must be transferred to the opposite side of the pivot point, for the cord cannot remain at a fixed length.[27] Fortunately, his analysis of tensions reveals how to deal with the resultant problem. Thus, the complete design for the clock (Fig. 3.11) shows a weight *A* moving in a circular fashion about the clock,

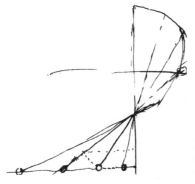

Figure 3.10. Preliminary drawing for the conical-pendulum clock with the bob at a constant height.

the cord passing from *A* over a pulley to a weight *E*, which has a chain hanging from it, with the links at the end stacking in a neat pile on a shelf. The weight *E* and that part of the chain *CD* that is suspended off the shelf act as a counterbalance to the tension on the cord exerted by the rotating bob *A* just as *L* counterbalances *H* in the geometric derivation.[28] Should the length of the cord from *A* to the vertex decrease because of some disturbance of the bob's rotation, the chain on the other side of the pulley will stack some of its links onto the reserve pile, thereby reducing the weight on that side and hence lowering the counterbalancing tension.

The preliminary drawing hints at what ensues, for it follows that *A* will be too heavy for its present position and will fall, reducing the angle made by the cord with the perpendicular. Presumably the bob will continue to fall inward until the angle is precisely that necessary for the tension and weight to balance once more. Although the bob will be closer to the vertical axis, the altitude of the cone it describes will be the same, and thus the time of revolution will be the same, ensuring that the clock keeps uniform time. In actuality, the process by which equilibrium is again achieved is far from simple, since the excess weight of the falling bob will be pulling on the cord, raising the weight *E* and increasing the number of hanging links, thereby increasing the tension on the cord.

In the process of designing this clock, Huygens calculates the lengths of cord necessary for a conical pendulum to complete 5,040 and 4,320 revolutions per hour, where the cord is assumed to remain at a 45° angle with respect to the axis.[29] Using his latest value for the distance fallen by a body in 1 second, namely 13 feet 8 inches, he finds that the lengths should be approximately 6 and 8½ inches,

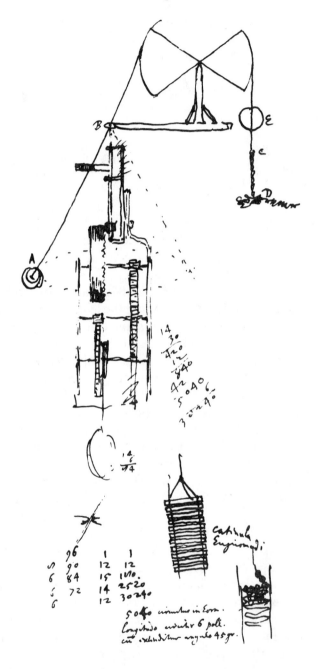

Figure 3.11. The conical-pendulum clock with its adjusting chain.

respectively.[30] The first length corresponds to a conical clock that rotates at slightly faster than ¾ second (0.72 second), whereas the second length yields a somewhat slower rotation (0.83 second).

These calculations and the sketches that they accompany are the only extant evidence from 1659 of a clock regulated by a conical pendulum. It is clear, however, that sometime during the first two weeks of November Huygens must have actually constructed and tested a conical-pendulum clock, thereby discovering that his value for the constant of gravitational acceleration was wrong; for by November 15, he possessed a new value for free fall – 8 feet 9½ inches in ¾ second – that he says he determined "from the movement of a conical pendulum" (*ex motu conico penduli*).[31] To obtain this new value from the foregoing parameters Huygens would have had either to lengthen the cord until the clock advanced the proper number of revolutions per hour or to count the actual number of revolutions circuited in an hour for a given length, at which point he could then have calculated the value of the radius for a circle traveled in ¾ second and applied his formula to determine the corresponding distance traversed by a falling body.[32]

On November 15, Huygens attempted to confirm his new value by repeating Mersenne's experiment for a third time. He used the pendulum of October 23, which supposedly took ¾ second to hit the wall when mounted for the experiment, a value that undoubtedly prompted him to choose gears for the conical clock that would make its pendulum rotate in roughly the same time. In addition, he tried to improve the method by which he performed the experiment. Instead of holding the bob of the pendulum and the loose ball in one hand, he connected them with a short string, which upon being cut would release the balls at exactly the same moment. The bob was made to hit parchment and the ball to fall into a lined box so that the moment of their striking would be more readily discerned. Huygens records that when the ball was dropped from 8 feet 7 inches it appeared to hit the box at exactly the same moment at which the bob hit the parchment. Unfortunately, he goes on to note, the two balls could be heard to hit simultaneously within a variance of 3 to 4 inches.[33] In particular, of course, this means that simultaneity would also occur if the ball were to fall from 8 feet 9½ inches.

Huygens clearly knew that Mersenne's experiment was faulty, even with all the improvements that he had made, but in his report of the retrial he concludes that the technique could at least corroborate the value achieved independently with the conical pendulum: "It suffices that the experiment not negate this measure, but to the extent that it can, it does confirm it." Although the record is ambiguous, it seems reasonable to assume that Huygens actually repeated Mersenne's experiment with the ball released from a height of 8 feet 9½ inches. Indeed, it is remarkable that he did not begin his repetition of the experiment by presuming this value was the correct result. Instead, his differing values clearly show that he maintained the integrity of the experiment as an independent method. Although there is very little evidence of his experiments and even less of his procedures (for example, repeated trials are not recorded, although he mentions them in this context), he obviously was a careful observer. The wide variation in results was the fault of the experiment, not of the experimenter. No wonder, then, that he abandoned Mersenne's experiment, crossing out the latest value of 8 feet 7 inches and replacing it with that obtained by the conical pendulum, 8 feet 9½ inches.

The new value leads to a constant of gravitational acceleration of *"proxime"* 15⁶⁄₁₀ feet (equivalent to 979 cm/sec^2) and even when the large variance is taken into account the value remains above 15 feet for the first second of fall.[34] This is a figure significantly different from his previous experimental values, which might explain why Huygens concludes his account of November 15 with a few more comments on the inadequacy of the experiment and a citation of Riccioli's 15-foot value, which does support the new result. Thus, by November 15, he had an empirically achieved value, extremely close to that accepted today, confirmed by an independent experiment and supported by evidence from another researcher.[35]

During the same period (November 5 to 15) in which he developed and tested his conical clock, Huygens had rewritten and expanded the theoretical justification of that clock, *De Vi Centrifuga*. The revision is readily dated because Huygens uses his value from October 23 (13 feet 8 inches) for the constant of gravitational acceleration in a proposition presenting his formula for the conversion of circular motion to free fall. Sometime after the revision's composition, Huygens crossed out the October 23 value and replaced it with that of November 15 (15⁶⁄₁₀ feet).[36]

The revised version of *De Vi Centrifuga*, as Huygens wrote it (as opposed to the treatise published posthumously under that title), differs very little from the preliminary draft. Some propositions are gone, including those referring to the suspension of a ball in chalices and on inclined planes, some are added, and those that are included have elegant geometric proofs attached. The primary addition is a preface discussing both the true and the approximate paths of centrifugal motion, including Huygens's famous relativistic scenario of a man rotating on a turning platform while holding a ball by a string. However, the insights discussed in this *De Vi Centrifuga* had already been suggested in the earlier draft, a fact that is not surprising, since the dates of composition of the two differ by as little as a few days and at most three weeks.

It is worth noting that the revised *De Vi Centrifuga* does not include Huygens's discovery of the isochronism of the paraboloid of revolution. In the earlier draft, he notes that a ball placed in a spinning chalice of paraboloidal shape transits all circles on the surface in equal times, but this proposition is not repeated in the redaction. The isochronism of the paraboloid does appear in the list of theorems appended to Part 5 of the *Horologium Oscillatorium* and consequently in the posthumous edition of *De Vi Centrifuga*, accompanied by a proof supplied by its seventeenth-century editors.[37]

That it should be so fated is understandable. Without a theory of evolutes, Huygens had no way to control the cord of a pendulum so that the bob would move on the imaginary surface of a paraboloid. Instead, he could work only with the nonisochronous conical pendulum, adjusting its tension so as to maintain the bob at a fixed altitude, thereby ensuring constant rotation. Later, with the theory of evolutes at his disposal, Huygens would have a means for developing a paraboloidal pendulum. Meanwhile, the isochronism of the paraboloid was of no practical value and was discarded. What remained, however, was a brilliant, detailed theory of centrifugal force and its physical model, the conical pendulum.

PRECURSORS

The extent of outside influence on this treatment of centrifugal force is difficult to assess. Certainly, Galileo and Descartes were deeply concerned with explaining circular motion, since it was fundamental to their cosmologies. Yet considering the importance of circular motion to their thought, their handling of the problem was remarkably vague, and their qualitative, sometimes fuzzy, explanations of circular motion allowed them to ignore basic flaws in their respective systems of the world. It must be admitted, however, that they had at least attempted an explanation and that Huygens knew and respected their work. What could he have drawn from them?

In the *Principia Philosophiae*, René Descartes discusses the circular motion of a stone in a sling with the intent of applying his findings to vortices. By his own principle of inertia, the stone would, if freed from the sling, move along a line tangent to the circle at the point of release, but it is constrained by the rotating sling to move instead along the circular path. The difference between the circular and tangential paths as measured along extended radii of the circle represents for Descartes that portion of the inertial motion of the stone that has been impeded by the rotating sling. From his drawings and examples, it is clear that Descartes envisions a pseudosummation of forces, the circular tendency plus the radial tendency equaling the inertial tendency. In accordance with this geometric view, Descartes argues that the radial tendency is continually increasing, but he never attempts to quantify this increase except to point out that as the speed of rotation increases so does the tension (*tensio*) exerted on the sling by the attempt of the stone to recede radially.[38]

Applying his analysis to vortices, Descartes argues that particles of matter endeavor to recede radially from the center of the spinning vortex but are prevented from such motion by the particles surrounding them, just as the stone is impeded by the sling. Actual movement, versus the tendency to motion, occurs only because the particles of matter are of various sizes and move at different speeds, and thus the pressures that are exerted on any one particle by its neighbors are uneven, allowing the particle to overcome its restraint.[39] Hence, although he has unequivocally stated the law of rectilinear inertia, Descartes has disallowed its actualization in his universe, since every particle is constantly being jostled and impeded by its neighbors in the spinning vortex. Indeed, because all particles participate in the circular motion of the vortex, only the radial component of the inertial tendency is perceived.

It is easy, too easy, to attribute Huygens's work on centrifugal force to an omnipotent Cartesian program. In such an interpretation, Huygens, always ready to accept the challenge of rendering Descartes's nebulous philosophy into precise mathematics, took up the problem of gravity in order to clarify and defend Cartesian vortices. Indeed, the passage on gravity that preceded Huygens's repetition of Mersenne's experiment can be interpreted as evidence for this view. The fact that a refined version of that discussion of gravity appears in the *Discours de la cause de la pesanteur* only substantiates Huygens's commitment to the Cartesian program.[40] Moreover, there are many superficial similarities between Descartes's and Huygens's discussions of centrifugal force. Both deal with the tendency to recede by citing the tension that centrifugal force exerts on a restraining cord or sling as proof of its existence and by using the difference between the circular and tangential paths to measure it.

Yet each of the points cited in favor of a strictly Cartesian interpretation of Huygens's work can be mitigated, if not negated, by a counterclaim. To begin with, by its nature, because it posits a spinning earth, Copernicanism turns the explanation of gravity into a problem of centrifugal force, one that every Copernican of the seventeenth century – the warrior for the faith, Galileo, as well as the closet believer, Descartes – had to address. Notwithstanding the differences in their philosophies, the tools these Copernicans used were often similar. Thus, although Huygens's pendulum cord is reminiscent of Descartes's sling, Huygens's description of a man on a spinning platform parallels Galileo's view of the earth as a spinning wheel. In addition, like Huygens and Descartes, Galileo measured the incipient centrifugal force by the difference between the circular and tangential paths.

Furthermore, only when Huygens abandoned his muddled Cartesian explanation of the causal relationship between centrifugal force and gravity in favor of an analysis of resultant tensions on a cord did he succeed in achieving his goal of measuring the constant of gravitational acceleration by means of the conical-pendulum clock. Although the tension felt when the pendulum is rotating in a horizontal plane might be attributable to Descartes, the tension felt when the pendulum hangs perpendicularly without swinging is strictly Galilean. Expressed in simplistic terms, Huygens's crucial insight was to balance the outward tendency of centrifugal force described by Descartes with the inward tendency of gravity studied by Galileo.

Galileo's treatment of circular motion is centered on the question of whether a spinning earth would eject its passengers. In the Second Day of the *Dialogo . . . sopra i due massimi sistemi del mondo,*

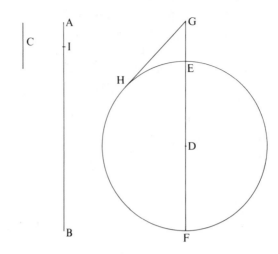

Figure 3.12. Galileo's secant proof: $HG/GE = AB/C$.

tolemaico, e copernicano he argues that the earth, viewed as a huge wheel, would never throw off a body on its surface:

For the distance traveled being so extremely small at the beginning of its separation (because of the infinite acuteness of the angle of contact), any tendency that would draw it back toward the center of the wheel, however small, would suffice to hold it on the circumference.[41]

He goes on to argue that the conclusion is valid even if the speed of rotation is very fast with respect to the speed of fall, as might be the case with a feather. For no matter how large the ratio between the distance that a body projected tangentially has moved and the distance that it must fall in the same time, a secant can be found such that the ratio between the tangent that the secant intercepts and the exterior segment of the secant is as large as the given ratio. Thus, in comparison with the tangential distance, the secant's segment is so very small that a body under a weak gravitational pull would still have time to reach the earth. Further, Galileo claims:

By degrees as the secant approaches the contact, this proportion becomes greater *ad infinitum*. So there is no danger, however fast the whirling and however slow the downward motion, that the feather (or even something lighter) will begin to rise up. For the tendency downward always exceeds the speed of projection.[42]

To reinforce his verbal argument, Galileo then rigorously proves that given any value *AB* (Fig. 3.12), arbitrarily large in comparison

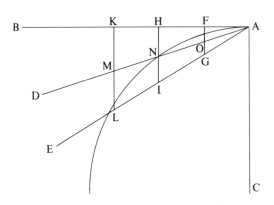

Figure 3.13. Galileo's comparison of fall with circular separation.

with another value C, a secant FG can be drawn to a given circle such that the tangent HG will have to the secant's segment GE the same ratio as AB to C. To make his construction Galileo finds the third proportional AI to AB and C, fixes G such that FE/GE equals BI/AI, and then draws the tangent from G to the circle. By construction, HG/GE equals AB/C.[43] Galileo is using the same secant relationship with which Huygens will begin his research on centrifugal force, namely $HG^2 = GE \cdot FG$. But instead of dealing with the exact formula at hand, Galileo moves on to the next part of his argument. Presumably, he feels that he has proved his entire claim.

In fact, he is wrong. Indeed, if his argument were true, the faster the spin of the earth (the larger the given ratio) the harder it would be to eject a body (the smaller the secant in relation to the tangent). His mathematical proof is valid, as far as it goes, but it does not justify his loose verbal conclusion that the proportion increases ad infinitum as the point of release is approached. The same kind of error undermines the next stage of his argument.

Galileo's interlocutor, Sagredo, asks whether the degrees of speed due to gravity on the body are not also becoming smaller as the secant approaches the contact point. To answer this question, Galileo introduces a diagram (Fig. 3.13) in which he marks off equal units of time along the tangential path BA and the velocity of fall in those times perpendicular to the units of time, FG, HI, and so on, supposedly showing thereby that the downward velocity diminishes more slowly than the separation of the tangent from the circle.[44]

With a great deal of hand waving and invoking ad infinitum, Galileo concludes, "Therefore let the tendency to downward motion be as small as you please, yet it will always be more than enough to get the moving body back to the circumference from which it is distant by the minimum distance, which is none at all."[45]

How similar and yet how different is Huygens's approach. At first glance, his initial diagram (see Fig. 3.1) is nothing more than Galileo's two drawings taken together: a tangent line and secants of the circle plus the parabolic path of the body falling under the influence of gravity. Admittedly, and significantly, Galileo did not draw the crucial parabolic path that represents distance with respect to time of fall, but rather graphed the linear relationship of velocity according to time. Yet the fundamental difference between Huygens's and Galileo's work does not lie there, nor even in the very apparent disparity of mathematical expertise.

The major difference lies in the two conceptualizations of the problem. For Galileo the goal was to prevent the body from leaving the surface of the earth, and the secant was a measure of the distance through which the body must return should it be thrown off. For Huygens the goal was to equalize the outward thrust of the body and the downward pull due to gravity, and the secant was a measure of that outward Cartesian force as well as a measure of the downward Galilean force. Galileo could never admit this equilibrium without having to concede that objects could fly off the spinning earth, given the right parameters; Galileo relied on the dominance of the inward tendency to keep the body in its circular path and thus to preserve his cosmology. For Huygens, less defensive about his Copernicanism, it sufficed to calculate with the actual parameters and to find that gravitational force is more than adequate (by a factor of 265) to keep a body from escaping into space. Huygens's study of centrifugal force was narrowly focused on the problem of measuring gravity by means of a conical pendulum, it should be remembered, and was not part of a general cosmological apology.[46]

In other passages in the *Dialogo* Galileo makes more qualitative statements regarding centrifugal force. At one stage, Galileo points to the universally accepted truth that a "body made to rotate by a single motive force will take a longer time to complete its circuit along a greater circle than along a lesser circle," citing as an example clocks with a balance regulator in which the time of vibration is shortened as weights are moved closer to the center of the balance bar (*foliot*).[47] Likewise, if two wheels of unequal size are rotated about their centers with equal speed, the smaller wheel is much more powerful at projecting a stone than the larger.[48] As the editors of Huygens's work have pointed out, Galileo's drawing for this argument is very similar to one that Huygens sketches when he states in a parallel proposition that the tension on a cord is in inverse proportion to the radius, when speed is kept constant.[49]

But Galileo's study of circular motion is always vitiated by his desire to prove that in all cases the spinning earth would retain its passengers. Thus, in the argument above, he concludes:

Since the casting off diminishes with the enlargement of the wheel, it might be true that to have the large wheel extrude things as does the small one, its speed would have to be increased as much as its diameter, which would be the case when their entire revolutions were finished in equal times. And thus it might be supposed that the whirling of the earth would no more suffice to throw off stones than would any other wheel, as small as you please, which rotated so slowly as to make but one revolution every twenty-four hours.[50]

Supposing the times of revolution of the two different-sized wheels are the same, Galileo has already argued in an earlier passage:

The velocity would now be greater in the large wheel merely by reason of its greater circumference. No one would suppose the cause for extrusion to increase in the ratio of the speed of its rim to that of the smaller wheel; that would be quite false.[51]

"Galileo was deceived," Huygens later declares regarding this muddled account of circular motion.[52] Yet despite his delusion concerning the primacy of gravity, Galileo had provided a stimulating analysis, one that others could build upon. Indeed, upon reading the *Dialogo*, the ever inquiring Mersenne attempted to clarify Galileo's arguments by an appeal to numerical examples.[53] In *Harmonie universelle* he begins his discussion of diurnal motion with a presentation of Galileo's two geometric propositions comparing fall and centrifugal tendency, then continues:

It is necessary to examine by means of numbers that which Galileo wished to demonstrate by means of lines, to wit, that the space that a body ejected by the daily motion of the earth ought to traverse in order to rejoin the surface is so small that there is more than enough time for it to arrive there.[54]

Taking a succession of small angles, each representing a slight rotation of the earth in a minimal amount of time, Mersenne discovers that the ratio between the distance that a body would fall in a given time and the distance that it would be from the earth, if it were released and moved uniformly along the tangent in the same amount of time, is constant. Each distance decreases proportionally according to the square of the times, and thus the ratio of distances does not increase ad infinitum as Galileo had argued but remains the same.[55] True, Mersenne grants, the ratio is very large, but it is not the largest possible, and he proceeds to show that with a change in parameters the ratio could be altered. In particular, if the earth were to turn twice as fast and at the same time the effect of gravity were only one-twelfth its actual value, the distance fallen in a given time would be less than the separation of the inertial path and the earth in that same time.[56]

We have thus shown that it is not true that, although one may increase the motion along the tangent and decrease that which moves it along the secant, the path that the weight ought to traverse in order to reach the circumference would be so small that whatever time span is chosen it would always be more than sufficient.[57]

Mersenne then proceeds to a discussion of the Galilean proposition concerning the effect on centrifugal force of a change in diameter or speed of the turning wheel, again attempting clarification by numerical example.

Mersenne achieved a remarkable analysis. In pinpointing and correcting Galileo's errors, Mersenne made the crucial break from the omnipotence of the circle, for he showed that, if the earth were to spin significantly faster, the centrifugal force on objects on its surface would overcome their weight. Admittedly, his conclusion sprang from dogged pursuit of specific results rather than from metaphysical conviction; but having shown that in certain instances centrifugal force exceeds gravitational force, Mersenne prepared the way for Huygens's study of that special situation in which the two forces exactly counterbalance each other. Huygens's creative step was to realize that, because the two forces can be made mathematically equivalent, if one wishes to study gravity one can study centrifugal force instead.

How amazing, then, to note the juxtaposition of Mersenne's analysis with his discussion of his experiment for finding the constant of gravitational acceleration. The propositions in *Harmonie*

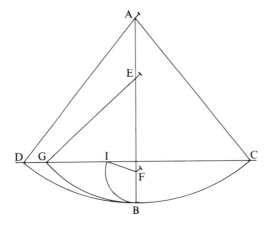

Figure 3.14. Galileo's drawing of a pendulum impeded by nails.

universelle that deal with centrifugal force follow immediately upon those in which he reports his attempts to use a pendulum to measure free fall.[58] Surely, it is only natural to conclude that the *Harmonie* was the source of inspiration for Huygens's work. Yet the seconds-pendulum that Huygens adopted in his first repetition of Mersenne's experiment was the pendulum of 3-foot length described in *Reflexiones*, not that of 3½ feet presented in *Harmonie*. Both works were in Huygens's library, however, and it seems very likely that, upon reading Mersenne's explanation in *Reflexiones* and encountering his own difficulties with the experiment, Huygens would have turned to Mersenne's other discussions of the topic.[59]

Likewise, once he undertook his study of centrifugal force, it seems probable that he would have remembered or reread the pertinent discussions by his two scientific masters, Descartes and Galileo. One proof that he was well aware of those previous attempts is his handling of Galileo's fundamental postulate on inclined planes, described in the *Discorsi*. In the process of justifying his postulate that a body will acquire equal speeds in equal heights when moving down differing inclined planes, Galileo introduces an ingenious example in which the inclined planes are replaced by arcs of circles of various radii. Consider a lead ball hanging by a thread from a nail *A* in the wall (Fig. 3.14). If the ball is set free at point *C*, its impetus at the low point *B* will be enough to carry it upward to the point *D*, which has the same vertical height as *C*. If a nail at point

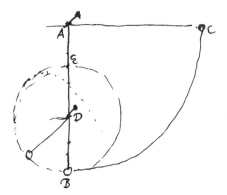

Figure 3.15. Huygens's drawing of
a pendulum impeded by a nail.

E or *F* now impedes the swing, the ball will still rise through the
arc *BG* or *BI* to the same vertical height.

If the interfering nail is so low that the thread advancing under it could
not get up to the height *CD*, as would happen when the nail was closer
to point *B* than to the intersection of *AB* with the horizontal *CD*, then
the thread will ride on the nail and wind itself around it.[60]

Huygens concludes the preliminary draft of *De Vi Centrifuga*
with a discussion that could only be a response to Galileo's distinctive example. He proves that if a nail is placed at *D* (Fig. 3.15),
where $DB = \frac{2}{5} AB$, the speed of the ball at the low point is sufficient to propel the ball about *D* with the cord kept taut, but that
if the nail is higher than *D* the ball cannot complete the circuit.[61]
As with so much of his work, Huygens had brought quantitative
precision to Galileo's physical insights.

These approaches to the problem of centrifugal force form a
progression of solutions. The first attempts to deal with the problem were the verbal, qualitative approaches of Descartes and Galileo. Although Galileo provided a semiquantitative solution, it was
Mersenne who attempted a numerical solution. Finally, Huygens
provided the mathematical theory, quantifying and structuring the
work of his predecessors. Certainly, Huygens's work went far beyond any of the three previous discussions, and however much he
might have gleaned from his precursors in the way of general conceptualizations of the problem, he felt no indebtedness, quoting
Horace on the first page of *De Vi Centrifuga:* "Freely I stepped
into the void, the first."[62]

Mathematical precision, in the sense of both the quantification
of a problem and the structuring of results, characterizes Huygens's

work, functioning as a creative, motivating element that pinpointed specific questions and carried him into new areas of research. Thus it was with Huygens's study of centrifugal force. Having associated gravity and centrifugal force, he proceeded very methodically from the two-dimensional geometric solution relating the two forces, to the three-dimensional description of physical reality in which both weight and centrifugal force act on a rotating object, through a variety of mechanical models corresponding to the three-dimensional geometric picture, to the ultimate choice of the conical pendulum, then to the mathematical analysis of that model culminating in the requirement that the pendulum maintain the same height regardless of the angle that it makes with the axis, and finally to the physical entity that embodied the foregoing theory.

This work constituted a brilliant, successful approach to what was in fact a rather narrow question, that of determining the constant of gravitational acceleration. By November 15, 1659, three weeks after he had begun, Huygens had solved that immediate, specific problem. He had achieved his experimental result by means of a conical-pendulum clock based on his theoretical studies and had corroborated that value with an independent test using Mersenne's technique.

Yet having achieved his experimental goal, Huygens did not end his research, but continued his pursuit of the theoretical consequences of his studies. He had measured gravity by transforming the question into a problem involving centrifugal force. Even in the midst of writing *De Vi Centrifuga*, however, he was trying to handle the question more directly, and soon after his successful completion of the indirect approach he returned to the mathematical equivalent of Mersenne's problem.

4

Accelerated motion: curvilinear fall

Mersenne and Riccioli had used the pendulum as a timepiece in order to measure the free fall of a body, but the pendulum is already per se a falling body, albeit restricted by the cord to move along a curve, and thus the swinging pendulum contains within itself the measure of gravity. Mersenne had dealt with this inherent relationship when he attempted to compare mathematically the fall of the pendulum's bob along the quarter arc of a circle with free fall through the length of the pendulum. As we have seen, Mersenne was not successful in handling the question, although he did at least achieve bounds on the ratio of times of fall.

With Mersenne's inadequate analysis as an example, Huygens also tried to quantify curvilinear fall during the period in which he was analyzing gravity by means of its analogy with centrifugal force. His first attempt, given in Propositions 19 and 20 of the first draft of *De Vi Centrifuga*, compares the hanging weight of the bob with the tension on the cord of the pendulum due to centrifugal force. The first of the two propositions states that a ball falling under the influence of gravity through the quarter arc of a circle exerts a tension at the bottom of its fall equal to three times its hanging weight. The second says that the tension on the cord after one complete circuit of an upright circle, in which the centrifugal force is precisely that necessary to keep the cord taut at the maximum height, is six times the hanging weight. Huygens's analysis of a pendulum impeded by a nail, drawn from Galileo's postulate, follows as a corollary to the latter proposition.[1]

Both propositions are obvious extensions of Huygens's study of centrifugal force to the situation of Mersenne's experiment, in which the circle, instead of being horizontally placed and considered gravity-free, now stands upright with both gravity and centrifugal force influencing the motion of the circuiting ball. Both results are derived from his basic formula relating centrifugal force and weight; that is, they are achieved by mathematically overlaying the circle and the parabola of fall. Yet both propositions, although interesting, brought Huygens no closer to the solution of the mathematical statement of Mersenne's problem. Thus, he undertook another approach.

Behind Mersenne's analysis of fall along the quarter arc of the circle stood Galileo's discussion of the same subject, in which he showed that fall along the circle is faster than along any chord or combination of chords and consequently concluded, erroneously, that the circle is the curve of fastest descent.[2] Galileo had tried to extend his results regarding fall along inclined planes to the circle by viewing the circle as an infinite collection of planes (its tangent lines), and although Mersenne was quite willing to adopt the idea of a curve as an infinite collection of planes, he was unable to deal with the concept mathematically. In addition, Galileo had used the fall of a pendulum to justify his basic postulate that a body acquires equal speeds when moving down inclined planes of equal heights but of differing lengths. Thus, a degree of circularity existed in Galileo's work. The fall of the pendulum explained fall on planes, yet fall on planes explained fall along a circle. Once again Huygens was confronted by an unresolved question with a partial solution and basic tools provided by Galileo and Mersenne; once again he would shape the problem into a structured theory far surpassing that of his predecessors.

THE GALILEAN TREATISE

During the period in which he composed *De Vi Centrifuga,* between the first and second drafts, Huygens wrote a treatise on fall, incorporating Galileo's work on inclined planes and attempting to expand it to curves.[3] As was his custom, he proceeded in a very orderly manner, stating all the salient propositions. He begins his summary by returning to the topics that had occupied him thirteen years earlier and had attracted Mersenne's attention to the fledgling scientist. After claiming that a body falls through all degrees of speed from rest, he argues that free fall engenders an accelerated motion that obeys the odd-numbers law.[4] In the first unit of time, he notes, a falling body attains a certain speed with which it would continue to move inertially were the pull of gravity removed; but

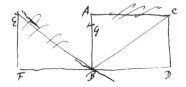

Figure 4.1. Huygens's version of Torricelli's principle.

in the next unit of time (and in all those subsequent) gravity is still acting, adding another unit of speed equal to the first. Thus, a falling body is uniformly accelerated, as Galileo claimed. Huygens then proceeds to a more exact proof of his statement to Mersenne that the spaces traversed in equal units of time are as the progression $1, 3, 5, 7, 9, \ldots$.[5] His second proposition is a variant of the mean speed theorem, which he first treated in 1654 while dealing with percussion. He shows that, if a body falling from rest traverses a given space in a certain time, the body would traverse twice that space in the same amount of time if it were moving with a uniform speed equal to the terminal speed of its fall.[6]

Rather than assume Galileo's postulate that bodies falling along various inclined planes reach the same speed at equal altitudes, Huygens attempts a proof based on mechanical principles. Reducing the postulate to an analogous problem in percussion, he begins by claiming that a perfectly hard body falling perpendicularly onto another, to which it communicates none of its own motion, rebounds to the same altitude from which it began and does so in the same amount of time. He then generalizes this idea to show that a body falling down an inclined plane acquires the same speed at the bottom as it would in falling freely through the height of the plane. Let a body fall along incline *BC* (Fig. 4.1), and assume that it has acquired less speed upon reaching *B* than it would have acquired in falling through *AB*. It would, nonetheless, acquire a speed equivalent to that attained by a body falling through a smaller space *GB*. By the very choice of *G*, if the body were to fall from *G* to *B*, rebound off a tilted plane placed at *B*, and then move up the plane *BC*, it would be able to travel the entire plane, arriving at point *C*. However, for a body to arrive at a height higher than whence it began, Huygens continues, violates Torricelli's mechanical principle that the center of gravity of a system cannot raise itself. Therefore, in falling along *CB*, the body must acquire a speed equal to that which is attained in falling freely from *A*. By the same argument, a body falling through any other inclined plane of the same height would acquire the same speed as it would down the given plane, and Galileo's postulate is thus proved, given Torricelli's principle.[7] This is followed by his rendering of another Galilean proposition showing that the times of descent on planes of the same height are proportional to the lengths of the planes.[8]

Turning to Galileo's restatement of his postulate in terms of the pendulum, Huygens generalizes the previous propositions to curvilinear paths. A body moving down any number of contiguous planes will acquire a final velocity equal to that which results by falling freely from the same initial altitude. Since a curve can be considered a collection of infinitely many contiguous planes (tangent lines), Huygens continues, the result also applies to the circumference of a circle or any other curve. Likewise extending his percussion-based proposition, he argues that a body falling through any curve will rise, be it along a curve or a plane, to the same height from which it fell. Of course, this is the fact that Galileo noticed regarding the motion of a pendulum and used in justifying his postulate. More generally, Huygens states, at any point in its fall or rise along a curve, a body will have the same speed as it would have by falling perpendicularly through a distance equal to its current vertical drop from the point at which it began its descent. Thus, a pendulum will ascend and descend equal arcs in equal times.[9]

All these propositions on fall are incorporated into the beginning of Part 2 of the *Horologium Oscillatorium*.[10] Huygens transferred them from his workbook almost wholly intact, although he provided some with proofs more truly geometric. However, even though they constitute a succinct treatise on fall, these propositions do very little more than make explicit what Galileo assumed intuitively.

The last two propositions are rather abortive efforts to move beyond Galileo's foundations. The first of them, similar to the work Huygens was doing at the same time in *De Vi Centrifuga,* introduces the mechanical concept of a sustaining force on an inclined plane. "The distances on differing inclined planes that are traversed by the same body in the same time, are in the same ratio as the forces [*potentiae*] by which they [*sic,* the body] can be sustained in each plane."[11] This proposition does not seem to have led Huygens to any fruitful results, and it is not included in the *Horologium Oscillatorium.*

The second is obviously an attempt to apply the formula for time of fall along inclined planes to curves, in a return to Mersenne's problem of comparing free fall with the swing of a pendulum. Huygens compares the time of fall along two sets of contiguous planes that are similarly arranged as to the angles of inclination and ratios of lengths of their members and concludes that the times of fall will be as the square roots of the lengths of the planes.[12] Unfortunately, he cannot simply generalize his result to curves by noting that they are composed of infinitely many contiguous planes, as he did in a previous proposition, because this proposition depends on a strict matching of each member of the set defining the first curve to one in the second set. Although a variation of this proposition generalized to curves would appear in the *Horologium Oscillatorium,* Huygens did not succeed in determining the time of fall along a curve during this study and proceeded no further.[13] Thus, his second attempt to handle the mathematical aspect of Mersenne's problem failed.

Immediately after completing this treatise on fall and just before designing his first conical clock, Huygens records the two salient features about time that he does know, writing in his workbook: "The times of vibration of pendulums are in subduplicate proportion to their lengths. The times of horizontal circulations are in subduplicate proportion to the heights of the cones that they describe."[14]

Less than a month later and two weeks after successfully handling Mersenne's problem experimentally by means of the conical pendulum, Huygens returned to the mathematical question of comparing fall along the quarter arc of the circle with free fall through the radius. Eschewing auxiliary devices such as inclined planes and cords under the tension of centrifugal force, his approach was a direct derivation involving only infinitesimals of length and time of fall. Once again he failed, confirming his own prediction: "The time through the quadrant of the circle is sought, but I doubt that it can be found."[15]

THE ISOCHRONISM OF THE CYCLOID

On December 1, 1659, Huygens posed himself a related, but significantly different problem. He must not have had much confidence that he would solve even it, for the question is written in the corner of a page already littered with other work and with little room for an extensive calculation (Fig. 4.2):

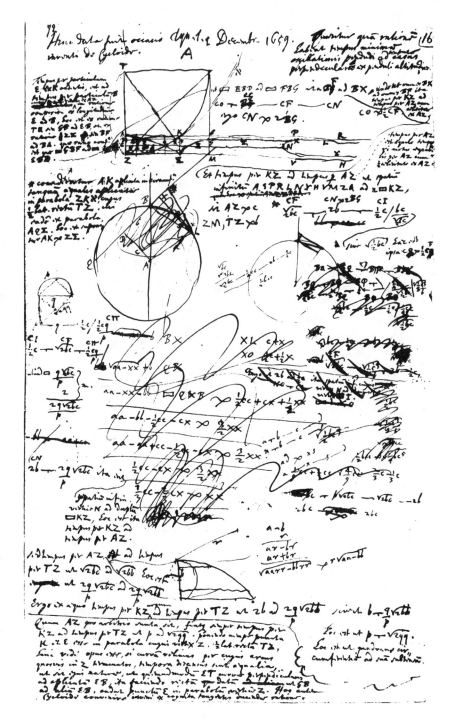

Figure 4.2. *Codex Hugeniorum* 26, f. 72r.

It is asked, what ratio does the time of a very small oscillation of a pendulum have to the time of perpenducular fall through the height of the pendulum.[16]

This shift from stating the problem for the whole quarter arc to viewing it at the infinitesimal level is crucial. The simple pendulum is not isochronous, as Mersenne himself knew but ignored in his calculations, and therefore the position at which the bob is released is important for determining the time of swing of the pendulum. By considering a minimal arc of swing, whereby the circle is approximately an isochrone, Huygens might have some hope of determining a relation for free fall independent of the arc length. Indeed, in this attempt he was successful at last. Moreover, through his consummate talent as a geometer, he was able to extend his solution to the cycloid, without the restriction to a minimal swing, and the result of his new investigation was his discovery of the isochronism of the cycloid.

This discovery is a direct product of Huygens's mathematics. To cite the result and accompany it with a brief modern derivation does nothing to trace or appreciate his creative act. Often his work has a self-evident quality; step 2 seems a "natural" continuation of step 1. Such was the case with the flow of ideas leading to *De Vi Centrifuga* and its companion clock. However, the isochronism of the cycloid was not one of those easy victories; Huygens resorted to a variety of mathematical manipulations in his attempt to transform the problem into something manageable, recognizable. Precisely because it was not a straightforward discovery, his methods are clearly exposed. Thus, although those who are not geometrically minded may find the following few pages a mathematical labyrinth, Huygens's own approach must be presented.[17]

Assume that the pendulum's bob begins its fall under the influence of gravity at an arbitrary point K on a circular path ZEK of small arc (Fig. 4.3).[18] Huygens intends to compare its fall with the fall from A of a body under uniform motion whose fixed speed is that which a body falling from A under the force of gravity would attain at Z. Thus, Huygens implicitly uses his terminal speed variant of the mean speed theorem to reduce vertical gravitational fall to the simpler case of uniform motion over the same path, whereby if v is the speed at point Z of an object falling under the influence of gravity starting from A, then a similar object moving from A

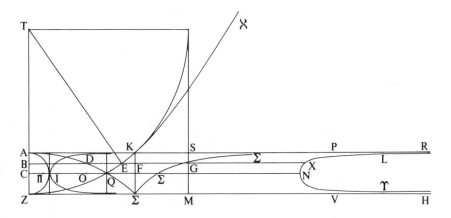

Figure 4.3. Huygens's isochronism derivation.

with uniform velocity of v will reach Z in half the time. Huygens deals with the factor of ½ later in his discussion.

Huygens begins his analysis by comparing the time of fall through the infinitesimal (*particula*) at E on the circular arc with the time through the infinitesimal at B on the vertical axis. By similar triangles, one having the infinitesimals at E and B as sides,[19]

$$\frac{\text{infinitesimal at } E}{\text{infinitesimal at } B} = \frac{TE}{BE}.$$

From his work generalizing Galileo's propositions regarding motion on an inclined plane, Huygens knows that at any point such as E on a curve the speed of a body under the action of gravity equals the speed of a body falling freely from the same initial height A to the same vertical height B. Since the vertical distance fallen AB is proportional to the square of the speed, this speed can be represented graphically (Fig. 4.3) by a parabola whose vertex is at the initial height A and whose base $Z\Sigma$ can be arbitrarily chosen equal to AK (because everything is expressed proportionally). The speed at each point E of the circular path of fall can thus be represented in any statement involving proportions of speed by the horizontal coordinate BD of the parabola $AD\Sigma$.[20] It follows that $Z\Sigma$ ($=AK$) represents the speed v acquired at Z by a uniformly accelerated fall from A and, therefore, also represents the uniform speed with which

Huygens assumes the body falling vertically will move along *AZ*, including along infinitesimal *B*.

This manipulation is quintessential geometric kinematics as practiced by Huygens. The necessity of representing accelerated motion geometrically led to the introduction of the parabola of fall, just as he had introduced it into his calculations of centrifugal force. Again he has finessed the physics into pure geometry. Although to a modern eye the construction (and similar geometric contortions that follow) appears very awkward, this conversion to geometric proportions allowed Huygens to avoid introducing the very factor that he did not know, namely the constant of gravitational acceleration that would permit him to express fall with an algebraic formula.

If motion over the infinitesimals at *E* and *B* is presumed constant, then the formula that time of travel equals the distance divided by the speed applies, and the parabolic representation of speed can be combined with the proportion for distance gained from the infinitesimal triangle to yield the opening statement of Huygens's derivation:

$$\frac{\text{time through infinitesimal } E}{\text{time through infinitesimal } B} = \frac{TE}{BE} \cdot \frac{Z\Sigma}{BD}.$$

Huygens then defines a new curve *LXN* that incorporates this new proportion by picking *X* such that

$$\frac{BX}{BF} = \frac{BG \cdot BF}{BE \cdot BD},$$

where *BF* is the same as *Z*Σ (the terminal velocity *v*) and *BG* equals *TE* (the radius of the circle). Inserting this new relation involving *BX* into his original proportion gives

$$\frac{\text{time through infinitesimal } E}{\text{time through } B \text{ with uniform speed}} = \frac{BX}{BF}.$$

Summing over all the infinitesimals, noting that the denominators are constant values, yields

$$\frac{\text{time through } KZ}{\text{time through } AZ \text{ with uniform speed}} = \frac{\text{all } BX}{\text{all } BF}$$

$$= \frac{\text{infinite space } APRXNHVZA}{AZ \cdot Z\Sigma},$$

where "infinite space" indicates an unbounded region with an area that is not necessarily infinite and $AZ \cdot Z\Sigma$ refers to the area of the rectangle $AK\Sigma Z$. By the mean speed theorem as previously applied, the time of fall through AZ of a body moving with uniform speed is one-half the time taken in falling through AZ under gravitational acceleration. Therefore,

$$\frac{\text{time of gravitational fall through } KZ}{\text{time of gravitational fall through } AZ} = \frac{\text{infinite space } APRXNHVZA}{2AZ \cdot Z\Sigma}.$$

At this point in his derivation, Huygens appears to hunt for some way to deal with this ratio of areas. He writes down two proportions involving segments of the line CN, where C is the midpoint between A and Z, but without any successful result. Then, once again resorting to a trusted technique, he assumes the equivalence of the infinitesimal arc of the circle, arc ZEK, with a parabola $ZE\aleph$ whose vertex is now at Z and whose *latus rectum* is $2TZ$. In particular, as he notes, this parabola is the inverted image of the parabola $AD\Sigma$ representing the speed BD, because $Z\Sigma$ has been chosen equal to AK. A series of algebraic manipulations follows, yielding another, quite significant proportion.

Fortunately, Huygens's sketchy calculations at this juncture can be understood with the aid of another manuscript (Fig. 4.4), in which he outlines without proofs the propositions that he has used in the main text.[21] The first of these propositions states that the speed of a falling body can be represented by a parabola, a fact already invoked in this reconstruction of his work, the parabola in question being $AD\Sigma$.[22]

The second proposition is a graphical representation of the mean speed theorem for falling bodies. A companion curve to a parabola is drawn (Fig. 4.5), using the mean proportion $BD/BK = BK/BL$ to define its points. Since BK is constant, BL, where L is on the new curve, is in inverse proportion to BD. Because BD represents the speed at the infinitesimal B of a body falling from A to B (first proposition) and time is inversely proportional to speed, Huygens views BL as the time of fall through the infinitesimal B. In particular, the time through the infinitesimal O with speed OF ($= BK$) is represented by OF. Again, the language of proportions allows him to introduce arbitrary factors without violation. Summing the times BL through all the infinitesimals on AO, which is his goal in the main text, gives the "infinite space" $OFHLZA$, which, according

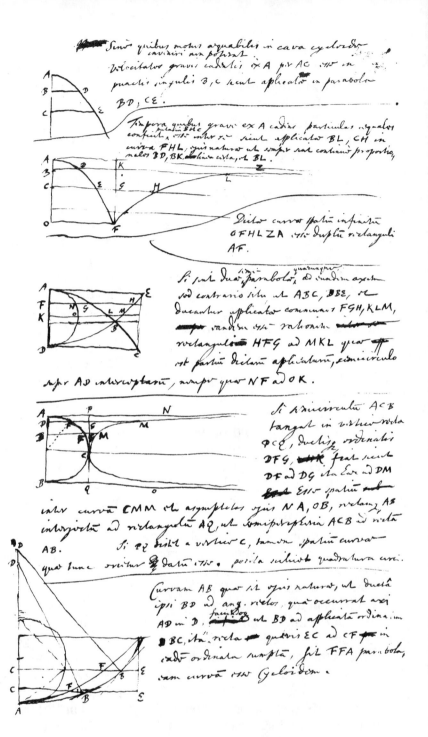

Figure 4.4. *Codex Hugeniorum 26*, f. 73r.

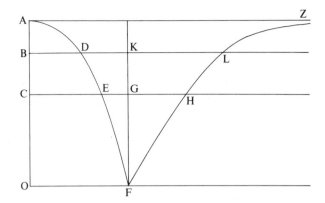

Figure 4.5. Huygens's variant of the mean speed theorem.

to Huygens, is double the rectangle AF $(= AO \cdot OF)$ formed by summing the fixed times OF through all the infinitesimals on AO. In other words, the time it takes a body to move from A to O under accelerated fall is equal to twice the time it would take the same body to traverse that distance moving with a uniform velocity equal to the terminal speed of the accelerated motion – which is Huygens's variant of the mean speed theorem. Huygens has drawn the curve representing time from this proposition into his main diagram (see $\Sigma\Sigma\Sigma$ in Fig. 4.3), probably an indication that, when he attempted to represent the time of fall through the arc of a circle geometrically, he first tried applying this proposition before he struck on a more fruitful approach using his next proposition.[23] Even the greatest of geometers was capable of a false start, and the proposition reflects his method of scanning his list of available techniques until the appropriate tool revealed itself.

Huygens's third proposition explains why he introduced the second parabola $ZE\aleph$ (Fig. 4.3) into his main derivation, for the proposition relates paired values on two parabolas to a corresponding sine of a circle. Returning to the curve LXN defined by the proportion $BX/BF = (BG \cdot BF)/(BE \cdot BD)$, note that both BF and BG are constant, and hence the definition of BX depends only on its inverse relationship to $BE \cdot BD$, where D is located on the parabola $AD\Sigma$ and E is on the circle ZEK. Thus, given two points X and x on the curve XNY, they would be related by the proportion $BX/bx = (be \cdot bd)/(BE \cdot BD)$; but if ZEK is now considered

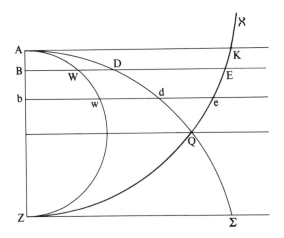

Figure 4.6. Introducing the second parabola (original lettering changed).

equivalent to the parabola *ZE*א (Fig. 4.6), Huygens can apply the relationship $(be \cdot bd)/(BE \cdot BD) = bw/BW$, where W and w are on the circle with diameter AZ.[24] Combining the two proportions gives $BX/bx = bw/BW$, and thus BX and BW are inversely proportional. Since Huygens's goal is to find the area of the space swept out by "all BX" in order to evaluate his proportion involving the times of fall, any reduction of BX to a known quantity, such as this reduction to the semichord of a circle, is a potential improvement.

However, in the language of proportions it would be even more advantageous if BX were directly, rather than inversely, related to a manageable quantity. Thus, having reduced the original proportion defining the curve *LXN* to a relationship involving a circle, Huygens next considers a second "infinite space" (Fig. 4.7) involving that same circle and based on still another mean proportion, namely $BW/BM = BM/BJ$, where BM is constant and J is on the new curve.[25] Because BJ is deliberately created as the inverse to BW, which is in turn inversely related to BX, this proportion will make BJ directly related to BX. Huygens claims in the fourth proposition of his auxiliary list that from the given proportion it follows that

$$\frac{\text{infinite space } APRJIHVZA}{AZ \cdot ZU} = \frac{\text{arc } AIZ}{AZ};$$

that is, the sum of all the BJs to the sum of all the BMs is as $\pi/2$. This result can be reconstructed by an appeal to the infinitesimal

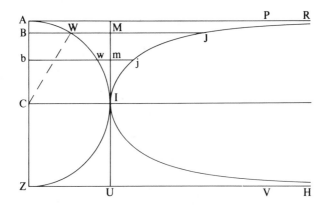

Figure 4.7. Introducing the second infinite space (original lettering changed).

triangle in a manner similar to that used by Huygens to begin his great derivation.[26] From similar triangles,

$$\frac{\text{infinitesimal at } W}{\text{infinitesimal at } B} = \frac{CW\,(=CI)}{BW} = \frac{BM}{BW},$$

and the defining proportion yields

$$\frac{\text{infinitesimal at } W}{\text{infinitesimal at } B} = \frac{BJ}{BM}.$$

Summing over the infinitesimals gives

$$\frac{\text{arc } AIZ}{AZ} = \frac{\text{all } BJ}{\text{all } BM} = \frac{\text{infinite space } APRJIHVZA}{AZ \cdot ZU}.$$

Still another proposition, recorded at a later date on a separate page, finally relates the original infinite space *APRXNHVZA* (all *BX*) and its rectangle *AZΣK* (all *BF*) to the second infinite space *APRJIHVZA* (all *BJ*) and its rectangle *AZU* (all *BM*).[27] For each point *B*, *BX/BJ* equals the constant *CN/CI*, where *C* is the mid-point of *AZ* (Fig. 4.3), because *BJ* was defined by the equation *BJ/CI = CI/BW* (since *CI = BM*), and in turn the definition of *BW* can be written *BX/CN = CI/BW*. Hence, (all *BX*)/(all *BJ*) equals *CN/CI*, which can be substituted into the composed proportion

$$\frac{APRXNHVZA}{AZ \cdot ZU} = \frac{APRXNHVZA}{APRJIHVZA} \cdot \frac{APRJIHVZA}{AZ \cdot ZU}$$

to yield

$$\frac{APRXNHVZA}{AZ \cdot ZU} = \frac{CN}{CI} \cdot \frac{CI}{C\Pi} = \frac{CN}{C\Pi},$$

where $C\Pi$ completes the proportion $CI/C\Pi = \text{arc } AIZ/AZ$ and, therefore,

$$\frac{APRXNHVZA}{AZ \cdot Z\Sigma} = \frac{APRXNHVZA}{AZ \cdot ZU} \cdot \frac{AZ \cdot ZU}{AZ \cdot Z\Sigma}$$

$$= \frac{CN}{C\Pi} \cdot \frac{C\Pi}{CO} = \frac{CN}{CO},$$

where CO completes the proportion $C\Pi/CO = ZU/Z\Sigma$.

The calculations that appear in the main text, immediately following Huygens's statement equating the circle ZEK with the parabola $ZE\aleph$, are the algebraic derivations of $C\Pi$ and CO. Taking $AZ = c$ and $TZ\ (= BG) = b$, Huygens has $CN = 2b$, $C\Pi = cq/2p$, where p/q is the ratio of the circumference of a semicircle to its diameter (that is, an algebraic equivalent of $\pi/2$), $Z\Sigma = \sqrt{2bc}$, and $CO = q\sqrt{2bc}/p$.[28] Thus, $CN/CO = 2b/(q\sqrt{2bc}/p)$. It follows that

$$\frac{\text{time through } KZ}{\text{time through } AZ} = \frac{\text{infinite space } APRXNHVZA}{2AZ \cdot Z\Sigma} = \frac{pb}{q\sqrt{2bc}}.$$

Since time in free fall is proportional to the square root of the distance,

$$\frac{\text{time through } AZ}{\text{time through } TZ} = \frac{\sqrt{c}}{\sqrt{b}},$$

and combining the two equations gives

$$\frac{\text{time through } KZ}{\text{time through } TZ} = \frac{pb}{q\sqrt{2bb}} = \frac{p}{q\sqrt{2}},$$

which is the ratio of one-fourth the circumference of a circle to its subtending chord ($\pi/2\sqrt{2}$). In other words, because the time through TZ is constant ($\sqrt{2b/g}$), the time of fall of the pendulum's bob from any arbitrary point K on the circle to point Z at the bottom is constant ($\pi\sqrt{b}/2\sqrt{g}$). Thus, the circular path on which K lies is isochronous, except for the fact that at one stage in the derivation it has been approximated by a parabola.

All that Huygens has achieved thus far is the proof of the fact already noted by Galileo that the circle is isochronous for small arcs. In this infinitesimal situation, he has found the exact value for the ratio of fall along the quarter arc of the circle to free fall, the ratio for which Mersenne had previously determined bounds. It was a belabored discovery, and anyone trying to retrace his path will readily agree to his later claim that it was not an easy achievement. Many of the manipulations of proportions strewn along the way were, in hindsight, unnecessary for the solution and, indeed, were questionable with regard to the validity of those "infinite spaces." He would eventually eliminate most of these proportions as extraneous, and the finished proof is nowhere near as tortuous or arduous to follow.

However, geometry was to reward him immediately for his persistence. The very same mastery that gave him the ability to advance until he hit upon the right combination of areas and proportions also granted him an unexpected prize. All those manipulations had not been in vain, for hidden away within the geometric substitutions were the requirements by which he could extend his result concerning isochronism to a full curve, not just an infinitesimal arc.

Reviewing his proof, Huygens realizes that his answer would be exact if K and E of the circular path ZEK (Fig. 4.3) were really on the parabola $ZE\aleph$ that he used as an approximation in his calculations. However, the opening step of his derivation, in which the infinitesimal triangle at E is replaced by a similar triangle TEB, requires TE to be normal to the curve of descent, a condition that will not be met if E is assumed to be on a parabola. Thus, the parabola will not be isochronous either.

Huygens needs a new curve (Fig. 4.8) that will preserve the proportion $DB/CB = CE/CF$, where B is on the new curve, DB is its normal, F is on the parabola, and CE is a fixed value. "However, this I found to fit the cycloid, by the known method of drawing its tangent."[29] With this simple statement Huygens announces his discovery of the isochronism of the cycloid.

Again, the auxiliary list of propositions helps to reconstruct the process by which Huygens might have made this claim.[30] As with the other propositions in this manuscript, he gives no explanation beyond the statement of the proposition and the accompanying diagram (see Fig. 4.4), in which, in addition to curves ABB and AFF, he has included a circle (labeled MPA in Fig. 4.8). Some faintly sketched lines depict the telltale property of the cycloid,

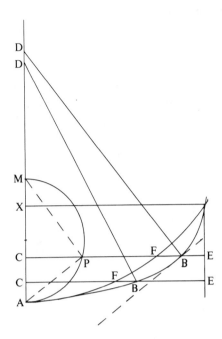

Figure 4.8. Discovering that the isochronous curve is a cycloid (lettering added).

namely that the tangent at any point *B* is parallel to a corresponding chord of its generating circle. This is the "known method of drawing a tangent" to the cycloid to which Huygens referred.[31] It should be noted that the circle is not the circle *AIZ* of the main text. Why should Huygens have introduced this circle?

The editors of his *Oeuvres complètes* begin their reconstruction of this proposition by assuming that Huygens knew that he wanted a cycloid and included its generating circle, choosing *MA* so that $CE^2 = pMA$, where *p* is the *latus rectum* of the parabola *AFF*.[32] In fact, for the proposition to hold, it is necessary that *MA* be the ordinate of *CE*, and in later proofs of the isochronism of the cycloid Huygens does include this requirement.[33] However, his drawing clearly shows that in this instance the ordinate to *CE* is not *MA* but another value left unmarked by Huygens (*XA*). Indeed, the inaccuracies of the drawing seem to indicate that he did not know *ABB* is a cycloid when he sketched the curve; for example, the normals should cross.

How then could Huygens have discovered that *ABB* is a cycloid? He has the defining relation $DB/CB = CE/CF$, where *CE* is constant and *F* is on the parabola $CF^2 = pAC$. The mean proportional properties of a circle define another "parabola" involving AC as ordinate, namely, $MA/AP = AP/AC$, or $AP^2 = MA \cdot AC$. Perhaps Huygens drew into his diagram an arbitrary circle, knowing that he could thereby relate CF to a chord AP of the circle.[34] Certainly, this technique had proved successful earlier in his main derivation when he related a pair of parabolas, one inverted, to chords of a circle. Since both CE and MA are constants, he could have then realized that the appropriate choice of MA would give him $CE/CF = MA/AP$. But this proportion, when substituted into the defining proportion, immediately yields

$$\frac{DB}{CB} = \frac{CE}{CF} = \frac{MA}{AP} = \frac{AP}{AC},$$

and thus the triangles DBC, PAC, and MPC are similar and the tangent to the curve at B is parallel to AP, which is the familiar property of a cycloid's tangent, when the circle MA is the generating circle and B is on the cycloid. Thus, without ever specifically finding MA he could have reached his conclusion that the cycloid is a curve that preserves the required proportion.

Of course, Huygens could also have just made an intelligent guess that the required curve was a cycloid. Coincidentally, among the scribblings that already filled the paper when he took up the problem of fall was a drawing of a cycloid.[35] However, the drawing does not include any tangent lines to the cycloid, and its presence near the concluding calculations of the problem at hand is much more likely a fortuitous juxtaposition.

Once he had hit upon the cycloid as the correct isochronous curve, Huygens rapidly refined his proof, first eliminating the "infinite spaces," then eliminating the approximating parabolas and infinitesimals.[36] By December 15, 1659, less than two months after taking up Mersenne's problem, he had a demonstration that is similar to the one that would appear in the *Horologium Oscillatorium* of 1673, and he wrote above it, "Great matters not investigated by the men of genius among our forefathers."[37]

Because of its importance to his thought and its almost complete disregard by modern commentators, Huygens's derivation of the isochronism of the cycloid has been presented here as completely as possible. One of the bewildering facts of modern scholarship in the history of science is the relative neglect of the rich and accessible collection of papers left by Huygens. Too often only the completed treatises are consulted, and the errors in interpretation that can result are exemplified by an article by Piero Ariotti. He assumes, without giving any justification for his claim, that Huygens knew of the isochronism of the cycloid from Lord William Brouncker, and thus: "He does not have to discover the correct curve. He knows he has that in the cycloid. Rather, he must *show* that the cycloid *is* the correct curve."[38] Ariotti then proceeds to explain the series of propositions in Part 2 of the *Horologium Oscillatorium* by which Huygens moves from motion along Galileo's inclined planes to motion on a curve and thence to the isochronism of the cycloid. In fact, Huygens's original derivation of the isochronism of the cycloid did not follow that path at all, but rather a long and arduous route. Only after he had discovered the correct curve could Huygens create the tidy proof that appears in the *Horologium Oscillatorium*.

Huygens was adamant about his originality on just this point. Prior to the publication of the *Horologium Oscillatorium* in 1673, both Brouncker and Ignace Pardies published proofs of the isochronism of the cycloid, each knowing from Huygens that the cycloid was the requisite curve.[39] In a letter to Henry Oldenburg that criticizes Brouncker's latest proof (his third attempt sent to Huygens), Huygens complains:

However, in my opinion the matter does not concern Lord Brouncker's honor, since it is no great thing for him to have demonstrated a proposition already discovered, which some others have also done, such as Father Pardies. . . . The principal and most difficult thing in matters of geometry is to find them, as those who discover them know very well.[40]

The calculations that Huygens used to discover the cycloid tell a great deal about the power of his mathematics. Despite all the mythology that has arisen about Huygens's inability to understand the calculus, he is revealed as a master of infinitesimal analysis. Throughout his derivation he uses indefinitely small particles of time and space, including the technique of summing infinitely many line segments to achieve an area. Likewise, he casually uses the infinitesimal triangle that Leibniz, inspired by Pascal, is supposed to have introduced years later.

Yet the genius for geometry for which he is usually recognized is also readily apparent. The fact that he moved so easily from one geometric image to another is proof of his mastery. For example, the transition from the infinite space swept out by all *BX* (Fig. 4.3) to a circle and then to another infinite space is done without recourse to elaborate calculations. In reconstructing Huygens's process of thought, the modern commentator, less adept at such geometric insights, gratefully resorts to Huygens's own list of propositions necessary for his derivation. The mere existence of this list, moreover, is evidence of the rigor of his standards of proof. Once insight had achieved the goal, only a strict accounting of all steps taken would suffice for its justification.

The very demand of representing entities proportionally, so often viewed today as cumbersome and archaic, led Huygens to the crucial reduction of the ratio involving his two auxiliary parabolas to an equivalent ratio involving a circle. This introduction into his diagrams of a circle in order to reduce unknowns to proportions inherent between chords was one of his standard and most successful techniques. The claim that Huygens was restricted by these ancient methods becomes difficult to substantiate when measured against the complex and fruitful geometric relationships that he achieved using them.

Indeed, although it contradicts the established view that seventeenth-century geometry was incapable of handling contemporary problems and thus necessitated the rise of the calculus, Huygens's success in deriving the isochronism of the cycloid occurred precisely because of his geometric approach. His initial approximation of the circle by the parabola, for example, reflects this geometric insight. Moreover, the recognition that his resulting geometric requirements for an isochronous curve describe the unique relationship between the tangents of a cycloid and the chords of its generating circle, although easily obscured in an algebraic expression, can be directly perceived in a pictorial representation.

For the second time in his pursuit of a solution to Mersenne's problem, Huygens's mathematical abilities yielded an unexpected prize. Just as his study of circular motion revealed the isochronous character of the paraboloid of revolution, so his study of gravitational motion led him to discover the isochronous nature of the cycloid. Never had he consciously sought the isochronous curve, as innumerable commentators have claimed, but the result came readily from his approach. In both cases, however, before his information could be of any practical value he had to devise a way for the pendulum's bob to move along the isochronous curve; the following chapter will discuss his success with this new problem. Indeed, within a week he had surmounted that next hurdle and was busy designing a cycloidal clock, the forerunner of that presented in Part 1 of the *Horologium Oscillatorium*.

The conical clock was not forgotten, however. In deriving the isochronism of the cycloid, Huygens had shown that the ratio of the time of fall along a very small arc of the circle or along the cycloid to the lowest point is to the time of free fall through the length of the pendulum's cord as the ratio of a quarter of the circumference of a circle to its subtending chord, and consequently the time of a double oscillation (one complete swing to and fro) to the time of fall through the length is as the complete circumference to the chord. Shifting to an obvious variation of the problem, Huygens proceeds to compare the time in which a conical pendulum makes one very small revolution with the time of free fall through the length of its cord. Listing a series of equations involving proportions of time of transit through various lengths and circular paths, he calculates that the ratio of the time of a "minimal gyration" to the time of fall through half the cord's length is as the ratio of the circumference of a circle to its radius. When extended by the times-squared law to the full length of the cord, the time of "minimal gyration" to the time of fall through the cord is also in the same ratio as the circumference of a circle is to the chord subtending a quarter arc. Thus, a pendulum of given length takes the same time to rotate conically through a very small circle as it would to complete one very small swing to and fro.[41]

This result can, and probably did, facilitate the direct transfer of Huygens's formula for the distance fallen freely by a body in the time of one rotation of a conical pendulum to a formula for the distance fallen in the time of a double oscillation of a cycloidal or restricted simple pendulum. Huygens had already proved in *De Vi Centrifuga* that two conical pendulums of differing lengths would keep the same time if their heights as measured along the axis were the same. Consequently, a minimal circulation of a conical pendulum of length L would take the same time as a rotation of a pendulum kept at 45° whose altitude, and thus radius, is L. However, in the time that it would rotate on a circle of radius L, with a speed sufficient to make its centrifugal force equal to its weight, a body could fall through a distance of $2L\pi^2$, according to the fundamental formula derived in *De Vi Centrifuga*. Since the time of a double oscillation of a cycloidal or restricted simple pendulum equals that of a minimal rotation of a conical pendulum of the same length, it follows that during the time of one double swing of a cycloidal or restricted simple pendulum of length L a body could fall freely a distance of $2L\pi^2$. In other words, the formula for the conical pendulum can be transferred to the cycloidal or restricted simple pendulum, rendering precise the correspondence, previously noted by Huygens, that the time of rotation of a conical pendulum is proportional to the square root of the height of the cone and that the time of swing of a simple pendulum is proportional to the square root of its length. Put in modern algebraic form, in all cases the time of vibration can be expressed by the formula $t = 2\pi\sqrt{L/g}$, where L is either the length of the cycloidal or simple pendulum (assuming a small double oscillation of either one) or the height of the cone.[42]

This extension of Huygens's result to a formula involving the cycloidal or restricted simple pendulum is a conjectured reconstruction, meant to account for the fact that soon after his discovery of the cycloid's isochronism on December 1, 1659, Huygens applied his formula to either a cycloidal or a restricted simple pendulum in order to measure once more the constant of gravitational acceleration. Thus, it should be emphasized that, having solved the mathematical problem of comparing fall through a circular arc with free fall, Huygens once again pursued the experimental aspect of Mersenne's problem.

The evidence of the experimental work done on or soon after December 1, as well as the evidence of the application of the formula itself, is extremely scant, consisting only of a few calculations made on the reverse side of the paper on which Huygens discovered the isochronism of the cycloid (Fig. 4.9).[43] Using the constant of gravitational acceleration ($15\frac{6}{10}$ feet) that he had achieved on November 15 by means of the conical pendulum, Huygens recomputes the radius of the circle circuited in 1 second, which he converts to $9\frac{9}{20}$ inches.[44] But in another calculation on the same page, reversing the derivation, Huygens uses $9\frac{1}{2}$ inches and the more accurate approximation of pi, $\frac{355}{113}$, to achieve a value of 15 feet $7\frac{1}{2}$ inches for the constant of gravitational acceleration.[45] He inserts

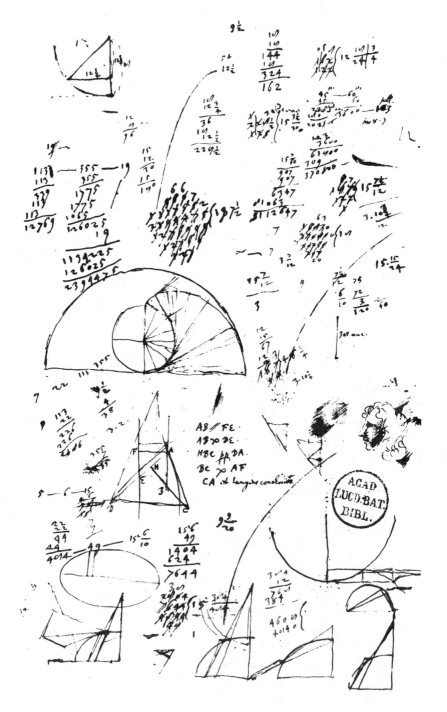

Figure 4.9. *Codex Hugeniorum* 26, f. 72v.

this new value into his report of the November 15 experiment and into the proposition of the revised *De Vi Centrifuga* in which he presents his basic formula relating free fall to motion in a circle, adding to the latter a marginal note to indicate that the constant of gravitational acceleration now stemmed from a "calculation" based on the December 1 results in which he compared free fall with the fall of the pendulum rather than on "experiment."[46]

The most likely interpretation that can account for the marginal note and avoid circularity in Huygens's calculations is one that assumes that $9\frac{9}{20}$ inches refers to the radius of the circle circuited in 1 second, whereas $9\frac{1}{2}$ inches refers to the length of a pendulum that makes a double oscillation in 1 second.[47] Consequently, the second derivation is evidence that Huygens had transferred his formula for the conical pendulum, equivalent to $g/2 = 2L\pi^2$, to the simple or cycloidal pendulum of length L. Now, because the same formula applies equally to the conical, the restricted simple, and the cycloidal pendulums of 1-second oscillation, once he had achieved the radius (L) of the circle circuited in 1 second, as indeed he had by November 15, Huygens theoretically possessed the length of a pendulum that completes one double oscillation per second. In other words, along with transferring the formula, Huygens could also have transferred the value of $9\frac{9}{20}$ inches to the simple or cycloidal pendulum. The discrepancy between the radius $(9\frac{9}{20})$ and the length $(9\frac{1}{2})$ indicates that the pendulum length must have been determined experimentally, even though an experiment was not necessary. The only reference to any such experiment appears in a letter from Huygens to Sir Robert Moray two years later:

I do not find it necessary to equalize the motion of the pendulum with cycloidal segments in order to determine this measure [the length of a pendulum that he wishes to make the universal measure]; rather it suffices to make it move with very small oscillations, which keep equal time close enough; and to seek thus the required length one needs to note, for example, a half-second[a] by means of a clock, which is already working well, and adjusted with the cycloid.

a) I have found this length to be exactly $9\frac{1}{2}$ inches of our Rhenish feet.[48]

It seems consonant with Huygens's methods of research that he would have attempted to verify by an independent means the theoretically achieved length, just as he had used his repetition of

Mersenne's experiment to verify the results of his conical pendulum experiment. However, even when it is presumed that an experiment was performed in December of 1659, there remains the question of whether it was the same as that described to Moray, and in particular whether Huygens used a cycloidal or a restricted simple pendulum. The traditional interpretation has decided in favor of the simple pendulum, the letter to Moray often being quoted without the qualifying remark concerning the synchronization of the simple pendulum with a cycloidal-pendulum clock. Yet it could just as easily be argued that Huygens created the method sent to Moray as an alternative to measuring the length directly from the cycloidal pendulum, which, as the members of the Royal Society (and perhaps Huygens himself?) discovered, was difficult to do. Enough evidence does exist to show that by the beginning of 1660 Huygens was designing clocks regulated by pendulums whose bobs moved along cycloidal paths. Further, the pride that he exhibited over this invention lends credence to the argument that he would have at least attempted the measurement using the cycloidal pendulum.

Certainly, by the beginning of 1660 Huygens had accepted 9½ inches as the true length of the pendulum that beats in 1 second and 15 feet 7½ inches as the constant of gravitational acceleration. This change is manifested in the aforementioned marginal note to *De Vi Centrifuga,* in which Huygens rejects the earlier value of 15⁶⁄₁₀ feet achieved experimentally *ex motu conico penduli* and replaces it with a value calculated from his analysis of the pendulum. In other words, in the most likely scenario, sometime in the first week of December he measured the length of a cycloidal pendulum that makes a double oscillation in 1 second and transferred this new value to the conical pendulum by means of his theoretical discoveries that equated a double oscillation with a minimal rotation. Thus, as far as the presentation in *De Vi Centrifuga* is concerned, the new value was from a "calculation" and not an "experiment," although obviously it still ultimately sprang from an experiment, Koyré notwithstanding.

In fact, Huygens could have achieved the value of 15 feet 7½ inches on November 15 without recourse to the discoveries of December 1 that actually led to its derivation. That he did not results from his frequent mathematical practice of truncating calculations at the tenths position, for when the necessary computations are performed and the division is carried only as far as the tenths position, the November 15 experimental value is 187.5 inches, which becomes 15⁶⁄₁₀ feet when the remainder is dropped a second time after division by 12.[49] In his calculation of December, Huygens achieves 187.5 inches by rounding, but represents it as 15 feet 7½ inches, thereby retaining the remainder. The only real difference between the value obtained from the conical pendulum and that from the cycloidal (or simple) pendulum was the manner in which a remainder was handled.[50] Thus, ironically, Huygens's editors, who did not thoroughly report the intermediate values achieved by Huygens on November 15 because they felt that 15⁶⁄₁₀ feet was no more than a poor truncation of 15 feet 7½ inches (= 15.625 feet), were correct. But fortunately for the traces of his process of discovery that these figures disclose, Huygens did treat them as different values in the period from November 15 to December 1, 1659. However, there is no way to judge whether, when faced with these two experimentally valid values, Huygens opted for 9½ inches because he discovered his mathematical laxness in the earlier derivation (not extant) or because he trusted the cycloidal (or simple) pendulum over the conical. As with the other aspects of this second experimental determination of the constant of gravitational acceleration, the sketchy manuscript evidence frustrates efforts to substantiate any conclusion.

His newly acquired value of 15 feet 7½ inches of fall during the first second of descent is equivalent to the modern accepted constant of gravitational acceleration for his latitude, 981 cm/sec^2.[51] However, the constant on which Huygens fixed his attention from 1660 onward was the length of the pendulum that makes one double oscillation per second, be it simple or cycloidal. Reversing its experimental function, he no longer desired to measure it more accurately, but instead wished to institute it as the standard of all other measurements, the universal measure so ardently sought by the fellows of the Royal Society of London.[52]

Indeed, even before the beginning of the new year, Huygens had moved beyond not only Mersenne's method for finding the constant of gravitational acceleration but the problem itself. Because both the centrifugal force on a ball being retained by a cord and the pull of gravity engender accelerated motions, Huygens had sought the solution of the problem in the uniform motion of a body on a circle. Through a careful analysis, the success of which depended on his manipulation of various cases in which the weight of a body is balanced by its centrifugal force, he was able to develop the fundamental principles of a theory of circular motion and its physical model, the conical pendulum. Using those tools, he was able to determine accurately for the first time the distance traversed by a freely falling body. He had the answer to the problem that he had set out to solve, and in addition he had discovered that an object moving on the surface of a paraboloid completes its circuit in a fixed time, irrespective of the altitude at which it is placed.

He had his answer, yet he continued to extend Mersenne's work, turning to the mathematical relationship embodied in Mersenne's experiment. What ratio does the time of fall of a pendulum moving through the quarter arc of a circle hold to the time of free fall through its radius? Putting aside the question as too difficult, Huygens studied its infinitesimal counterpart instead, finding that the time of fall along a very small arc at the bottom of the circle to the time of free fall through its radius is constant. His derivation led him to the discovery that, if the curve of fall is not a circle but a cycloid, the ratio holds regardless of the size of arc taken. The cycloid, like the paraboloid, is isochronous.

In solving his initial problem, Huygens had created a new one, to which he immediately turned after his triumph of December 1. How can a pendulum's bob be constrained to move along a cycloidal path or to revolve on a paraboloidal surface? Before he could apply his theoretical knowledge of isochronism to clocks, he had to answer this question. The solution led to his development of the theory of evolutes.

5

Evolutes

Because the bob of a freely swinging pendulum traces out a circular arc, which is not isochronous, Huygens had to adjust its swing in such a manner that the pendulum could beat equal times irrespective of the amplitude of its swing. The problem was not new to him. In 1657, when he first applied the simple pendulum to a clock, Huygens was fully aware of its nonisochronous nature for swings of large amplitude; and although at that time he did not know the path that the bob should follow in order to achieve isochronism, he was able to compensate for the irregularity of the simple pendulum by confining its swing between two curved metal bands whose shape he determined empirically. As he explained later to Pierre Petit:

In a simple pendulum the swings that are elongated more from the perpendicular are slower than the others. And so in order to correct this defect (the opposite to that which you believed) at first I suspended the pendulum between two curved plates..., which by experiment I learned in what way and how to bend in order to equalize the larger and smaller swings. And I remember having so well adjusted two clocks in this manner, that in three days they never showed between them a difference of even seconds: although in the meantime I often changed their weights, rendering them heavier or lighter. However, later because I found that with these plates the slightest tilt of the clock altered the length of the pendulum, I got rid of them, at the same time trying to make the vibrations of the pendulum narrower by means of the gears.[1]

A clock fitted with curved plates and dating from 1657 is pictured in Huygens's collected works.[2]

Although their shape was determined "by experiment," the plates did have a theoretical justification, evident in a brief manuscript that consists of some calculations and a drawing (Fig. 5.1) in which the pendulum's cord *HG* is restricted in its swing at points *F* and

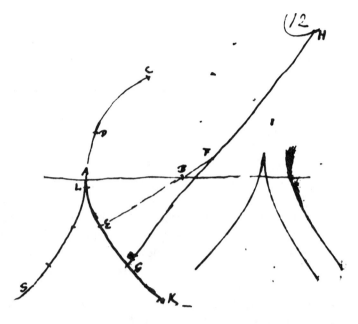

Figure 5.1. The restricted pendulum of 1657.

$B.$[3] In effect, the restrictions successively shorten the cord so that the bob sweeps out circular arcs EG and AE with correspondingly smaller radii EF and AB. As Huygens points out to Petit elsewhere in the same letter, Mersenne and Galileo had found that the time of swing of a pendulum is dependent upon its length.[4] Thus, this shortening of the effective length of the pendulum means that the bob will move through the arcs AE and EG more quickly than it would through an equal angular deflection with no impediment to its swing. Therefore, as the pendulum swings wider, its cord shortens and its speed of swing increases accordingly, with the result that the wider swing is completed in about the same time as a smaller, unconstrained swing. This pointwise restraint of the freely swinging cord is extremely reminiscent of Galileo's postulate in which the cord of a pendulum is blocked by a nail, forcing the bob along a different circular arc, and which Huygens would treat in *De Vi Centrifuga* two years later.[5]

In the case of the pendulum sketched in the manuscript, the path of the bob is a series of circular arcs, with points H, F, and B acting as centers of rotation. It should be noted that the cord at each instant of its swing is perpendicular to the arc being swept out, since a radius always meets (the tangents to) the circle at right angles. The shape of the deflecting plates for such a path would not be a smooth curve, but rather piecewise linear, with vertices at *H*, *F*, and *B*.

Huygens probably created the curved plates of the actual clock by bending metal bands in such a way that, when matched to the drawing (or one similar), they passed through the three specified points. Every point on the curved plates would then act as a deflecting point, and thus an instantaneous center of rotation or curvature. Therefore, implicit in Huygens's clock of 1657 is the very property that today defines an evolute (the plates), namely the locus of centers of curvature of a given curve (the bob's path).

However, empirically derived and inaccurate when tilted, the curved plates were abandoned by Huygens in favor of the simpler technique of restricting the pendulum to small, approximately isochronous oscillations through his choice of gears. It is this type of clock that is described in his *Horologium* of 1658. Thus, Huygens did not pursue the mathematical theory embodied in the plates of 1657, which could have led him to an analysis of evolutes. A great many commentators have claimed that the theoretical work of 1659 was undertaken to justify the experimental result of 1657, in an unbroken development of the basic idea.[6] On the contrary, for two years Huygens abandoned the curved plates of his early clock.

Nor did he make any attempt to trace the theoretical consequences when confronted with his second example of an evolute, that which arose during his work for *De Vi Centrifuga*. In their analyses of centrifugal force, Galileo and Mersenne had pictured the earth as a very large wheel, and in the revised *De Vi Centrifuga* Huygens extends this analogy by depicting the earth as a very large turning platform on whose edge stands a man holding a ball by a string. As the platform spins uniformly in gravity-free space, the ball attempts to follow the tangential inertial path. From the man's vantage point the ball, tugging on its leash, tries to move away radially. Although Galileo, Descartes, and Mersenne discuss this radial tendency, to which Huygens gave the name "centrifugal force," only Huygens explicitly notes that, in fact, if the ball is released from its restraint at *B* (Fig. 5.2), then when the man reaches *E*, the ball will not arrive at the intersection *C* of the secant and tangent lines but instead will be at a point *K* slightly toward the point of release *B*; likewise, when the man is at *F*, the ball will be at *L*, not *D*. Because the man and ball are moving with uniform speed, when the

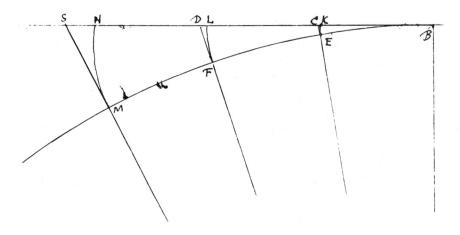

Figure 5.2. Turning platform, point of release fixed at *B*.

ball is released at *B* it moves with that same uniform speed along the tangent, and its distance from *B* at any time, therefore, is precisely the arc length of the circle between the man and the point of release. Thus, *BK* equals the arc *EB* and *BL* equals arc *FB*.

In order to trace, from the man's point of view, the ball's complete path of which *FL* is part, as is *EK*, Huygens adopts the relativistic approach of treating the man's position as fixed at *B* while the point of release moves away instead. Thus pictured (Fig. 5.3), the man no longer travels counterclockwise with a uniform speed; rather the point of release moves clockwise through points *L*, *N*, *M*, with the given uniform speed. As before, the ball's true position at any time can be located by measuring along the tangent line a distance equal to the length of the circular arc between the man and the point of release. Thus, when the man and the point of release are separated by arc *BN*, the ball will be at *R*, where *RN* equals arc *BN*. At a later time, when the point of release is at *M*, the ball will be at *S*, where *SM* equals arc *BM*.

If a thread is superimposed along the circumference with its end placed at the man's position *B* and if the thread is held fixed at the point of release, for example, at *N*, while the free portion toward *B* is straightened and laid along the tangent at *N*, then the end of the thread will mark the point *R* to which the ball will have traveled. To trace the entire curve *BRS* one need only carefully unwind

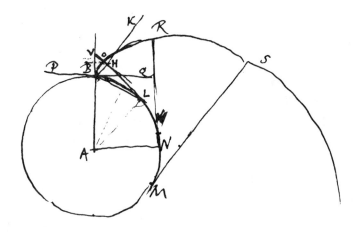

Figure 5.3. Turning platform, man's position fixed at *B*.

the thread off the circle, keeping it taut to ensure that the free portion forms a tangent.[7] In this way, the requirement that the ball's position along the tangent equal the corresponding arc length of the circle is given a mechanical derivation by Huygens that is precisely the technique that he would later use to draw the involute. Yet as with the curved plates of the 1657 clock, Huygens did not immediately pursue the mathematical consequences of what he had sketched.

A month later, however, having discovered the isochronism of the cycloid, Huygens at last concentrated his attention on this process of unrolling curves. Indeed, the reverse of the page on which he develops the isochronism proof contains a curious figure (see Fig. 4.9, midpage) in which the path from *De Vi Centrifuga* is surrounded by a cycloid, or perhaps a circle, as if he were trying to relate the curves. Certainly, the drawing suggests that Huygens immediately recalled the mechanism of unwrapping a cord from around a curved surface when he turned from his discovery of the isochronism of the cycloid to its application to clocks.

As it swung to one side, the pendulum of 1657 wrapped itself along the curved surface of a plate, while the free end of the cord remained straight and taut, perpendicular to the path of the bob and tangent to the curved surface at the point of contact. In other words, Huygens already knew a very basic relationship between the curve that the bob should follow and the curve that would define

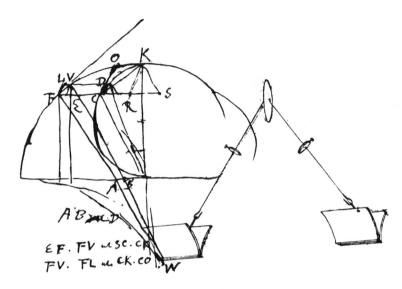

Figure 5.4. Deriving the evolute of the isochronous cycloid.

the plates, for the physical situation required that the normal to the bob's path be tangent to a plate. What he did not know until December 1, 1659, was the actual shape of either curve, for each had been empirically determined, with the bob's path defined by the unrolling of the pendulum's cord off the plates. The discovery that the bob should follow a cycloidal path gave hope that the shape of the plates could also be derived with mathematical precision rather than by experimental guesswork.[8]

Given two points, F and V, infinitesimally close on the cycloid FVK (Fig. 5.4), the normals to the cycloid through those points intersect at a point W that can be presumed to lie on the curve to whose shape the plates should be bent.[9] In order for Huygens to determine the point W geometrically, he must find the ratio FW/AW, which by similar triangles is the same as FE/AB. From the fact that the cycloid is constructed by rolling its generating circle along a baseline, AB is equal to the circular arc CD, which can be assumed equal to an infinitesimal segment of the tangent at C to the generating circle, since FV and CD are very small. Likewise, the infinitesimal FV of the cycloid can be considered equivalent to the tangent at F and thus parallel to CK by another property of the cycloid. Construct FL parallel and equal to CD and extend CD until it intersects at O the tangent at K. By their construction, triangles *FLV* and *COK* are similar, and since it can be shown that *COK* is isosceles, *FLV* is also isosceles, *FL* equaling *LV*.[10] Now, construct the point S on the extension of the line *FEC* so that triangles *FVE* and *CKS* are similar. In particular, therefore, angles *FVE* and *CKS* are right-angled, because *VE* is presumed normal to *FV*, and thus *CKS* can be considered as inscribed in a semicircle with diameter *CS*.[11] If R is the midpoint of *CS*, then the radii *RK*, *CR*, and *RS* are equal, making *CRK* an isosceles triangle. Thus, triangle *CRK* equals triangle *COK* with *CK* as their common base, and both are similar to triangle *FLV*. Since *CS* equals 2*CR* and thus equals 2*CO*, it follows that *FE* equals 2*FL* and thus equals 2*CD* or 2*AB*. Consequently, *FW/AW*, which equals *FE/AB*, is as 2 to 1, from which Huygens is able to conclude immediately that W lies on another cycloid. This conclusion is not readily apparent, but within the previous year Huygens had worked on precisely the problem that could help him at this stage. Just as he had the concept of curved plates at hand, so his knowledge of the cycloid was sufficiently developed for him to draw the necessary conclusion.

THE RECTIFICATION OF THE CYCLOID

In 1658, Blaise Pascal had challenged the mathematical community to solve a group of problems regarding the cycloid, including finding its area, center of gravity, and the volume and center of gravity of the solid obtained by revolving the cycloid about its axis.[12] Although the contest dissolved into acrimonious squabbling when Pascal refused to grant the prize to any of the competitors, the challenge did focus attention on this important curve and, in particular, prompted Christopher Wren to reveal that he had rectified (found the line equal in length to the arc of) the cycloid. When, shortly thereafter, Pascal wrote his *Histoire de la roulette,* an account of the competition and his solutions, he included a brief announcement of Wren's discovery:

However, among all the letters that I have received concerning this matter, there is none more beautiful than that which has been sent by M. Wren; for along with the pretty way in which he finds the area of the cycloid, he has given the comparison of this curve and its parts with the straight line. His proposition is that the length of a cycloid is four times its axis, the announcement of which he has sent without proof.[13]

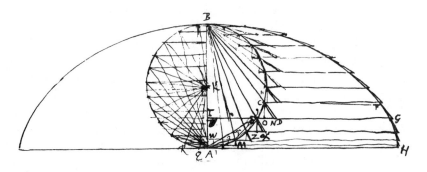

Figure 5.5. Rectifying segment *BG* of the cycloid.

Receiving a Latin edition of the *Histoire* in early January 1659, Huygens was greatly impressed by Wren's result, which constituted the first published direct rectification of a curvilinear figure.[14] Writing to John Wallis at the end of January, he describes Wren's determination of the arc length of the cycloid as "marvelously pleasing. It is the best result that has been discovered concerning this curve, for I distinguish the difficult from the elegant." Not content merely to admire, Huygens had immediately upon its receipt sought to verify Wren's conclusion, and thus having praised Wren, Huygens continues to Wallis, "I have discovered a general demonstration."[15]

Indeed, Huygens had successfully proved a more extensive proposition, one that determines the length of an arbitrary segment *BG* of the complete cycloid *BGH* (Fig. 5.5). If a parallel to the base is drawn through *G*, so that it cuts the generating circle at *E* and the axis *BA* at *T*, Huygens shows that $BG = 2BE$ and, as a consequence, $BH = 2BA$ and the complete cycloidal arc is four times the diameter. In the process, he determines the ratio of the arc *GH* to *AT* by an ingenious application of an Archimedean theorem concerning the sines of a circle.[16]

Huygens begins his demonstration by inscribing in the generating circle an equilateral polygon that has "infinitely many" sides and that includes points *A* and *E* as vertices. When parallels to the base are drawn to the cycloid from all the vertices contained in the arc *AE*, the sum of the infinitesimal segments of the cycloid's tangents cut by those parallels can be assumed equal to the length of the cycloidal arc *GH*. By the "well-known" property of the cycloid already encountered in the reconstruction of the isochronism proof,

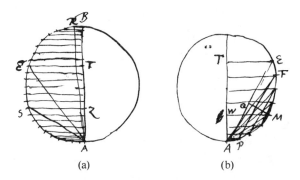

(a)　　　　　　　　(b)

Figure 5.6. Applying Archimedes' theorem on sines: (a) the theorem, (b) its application.

each tangential segment is parallel to the extended chord passing from the apex *B* of the cycloid to the corresponding vertex. For example, the tangent at *G* is parallel to *BEX*, and thus *EX* equals the tangential segment at *G*. Therefore, the ratio of arc *GH* (equal to the sum of all the infinitesimal tangential segments) to *AT* equals the ratio of the sum of all infinitesimals like *EX* to the sum of all infinitesimals like *EZ*, where *EZ* is the corresponding perpendicular segment of the parallel to *AT* at *E*. By the similarity of triangles *EZX* and *ETA*, these two infinite sums can be replaced, respectively, by "all chords under *AE*," indicating the sum of the lines that join *A* to all the vertices of the polygon under the arc *AE*, and by "all sines under *AE*," indicating the sum of the lines that fall perpendicularly from those vertices to the axis *AB* (lines such as *ET* and *MW*). Hence,

$$\frac{GH}{AT} = \frac{\text{all chords under } AE}{\text{all sines under } AE}.$$

Huygens follows this preliminary proportion with a second based on the Archimedean theorem, revealing what he knows about sines under an arc. The theorem states that twice the sum of the sines under *AE* (but not including *ET*) plus *ET* is to *AT* as *AR* is to *RB*, where *R* is the vertex of the inscribed polygon closest to *B* (Fig. 5.6a).[17] In fact, Huygens erroneously forgets the factor of 2, but because he immediately applies the theorem to a proportion and because that proportion involves a polygon with infinitely many sides, his restatement is valid. His version of Archimedes' theorem says that the ratio of all sines under AE to AT equals the ratio of AR to RB. The theorem applies to any arc on the circle; therefore, the ratio of all sines under AS to AZ equals the same ratio AR/RB. Huygens's second proportion follows:

$$\frac{\text{all sines under } AE}{\text{all sines under } AS} = \frac{AT}{AZ}.$$

If he could only change the chords of his first proportion into sines, he could use this variant of Archimedes' theorem to reduce those infinite sums signified by the word "all" to a ratio of simple lengths.

It is not surprising, therefore, that his next manipulation is a transformation of "all chords under AE" into "all sines under AE." Huygens claims that the sum of all the chords to arc AE from A equals twice the sum of all parallels to AE, such as PF, joining pairs of vertices under arc AE (Fig. 5.6b). The conclusion can be visualized thus: Consider an arbitrary chord AF; rotate F halfway toward E; likewise rotate A halfway toward P. The line will now be parallel to AE and, because it spans the same sized arc on the circle as before, its length has been preserved. Every chord from A connected to a vertex that is an even number of vertices away from E will, when rotated, connect a pair of vertices. The other chords (including AF), when rotated, will fall midway between lines connecting vertices, but can be considered in the infinitesimal situation to be equal to their immediate neighbors. Thus, each parallel to AE passing through the vertices will be covered twice.

In turn, however, each parallel is twice a sine under the arc EM, where M is the midpoint of arc AE and OM is the perpendicular baseline. Picture the whole system of parallels being rotated clockwise until E covers M, and it becomes apparent that the sines under arc EM overlay the sines under arc AM with baseline AW; in particular, $EO = MW$. Thus, with a very well developed geometric vision, one can easily see that the sum of all chords under AE equals four times the sum of all sines under AM. Therefore, the original proportion can be reduced:

$$\frac{GH}{AT} = \frac{\text{all chords under } AE}{\text{all sines under } AE} = 4\frac{\text{all sines under } AM}{\text{all sines under } AE}.$$

By Huygens's variant of the Archimedean theorem,

$$\frac{\text{all sines under } AM}{\text{all sines under } AE} = \frac{AW}{AT}.$$

Consequently, $GH/AT = 4(AW/AT)$, and $GH = 4AW$. In other words, the length of a cycloidal segment from any point G to the base of the cycloid is equal to four times the vertical drop from the midpoint M on the generating circle between G's corresponding point E and the base of the circle A. Since the result applies to the semicycloid, BGH equals $4AK$, or $2BA$. Wren's result follows as a corollary: The full cycloidal arc equals twice BGH, or four times the diameter BA.

To find the length of the segment BG, Huygens subtracts GH from BGH, which yields $2BA - 4AW$. Thus, $\frac{1}{2}BG = BA - 2AW$, but the latter value, according to a lemma Huygens provides, is equal to chord BE and, therefore, $\frac{1}{2}BG = BE$.[18] Consequently, the length of a cycloidal segment measured from the apex of the cycloid B to an arbitrary point G is equal to twice the chord drawn from B to G's corresponding point E on the generating circle. If one remembers that BE is also parallel to the tangent of the cycloid at G, the beauty of this result is apparent.

Huygens's proof for the rectification of the cycloid parallels the format used in his discovery of the isochronism of the cycloid. A problem that involves an infinitesimal triangle (that is, the derivative) is transformed by similar triangles to a noninfinitesimal situation involving chords and sines. Infinitesimals cannot be avoided, however, and once again Huygens is applying the theoretical technique of summing infinitely many line segments to achieve an area (the sum of the sines). But the brilliance of his discovery, as before, rests with his ability to manipulate geometric entities, to recognize, in this case, that the problem reduces to an application of Archimedes' theorem.[19]

Within a month of receiving Huygens's announcement of his generalized proof for the rectification of the cycloid, Wallis replied, stating that Wren had also achieved the length of an arbitrary segment of the cycloid and that Pascal had not fully reported Wren's findings.[20] Although Huygens was unable, therefore, to claim any priority for his discovery, his research made him cognizant of the general result almost a full year before the publication of Wren's proof, which Huygens did not receive until the spring of 1660.[21] Moreover, in achieving his own solution, he had attained that deeper familiarity that comes from working directly on a problem rather than merely reading another's proof. These differences in the depth and timing of his knowledge were crucial, for the rectification of

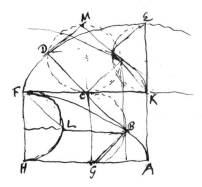

Figure 5.7. The cycloid's evolute is another cycloid.

an arbitrary cycloidal segment was precisely the element necessary for Huygens's successful discovery of the evolute of the cycloid.

THE EVOLUTE OF THE CYCLOID

Huygens's problem in December 1659, following his discovery that the cycloid satisfied the parameters necessary for isochronism, was to find another curve, the evolute, whose tangent at any point is perpendicular to the given cycloid and satisfies the relation $FW = 2AW$ (see Fig. 5.4). By the well-known property of a cycloid, the tangent at F is parallel to the corresponding chord CK on the generating circle, and thus the perpendicular FW is parallel to the chord on the circle that makes a right angle with CK; Huygens has sketched the chords parallel to FW and VW into his diagram. The chord corresponding to FW equals FA, or $\frac{1}{2}FW$, because they are parallel lines cut by parallels. Moreover, FW represents a cord that has been unwound off the desired curve, and thus it is the length of that evolute. What curve has an arc length equal to twice the length of a corresponding chord on a circle? Could the evolute of the cycloid be, by some fortuitous coincidence, another cycloid?

Apparently Huygens surmised as much, for in an extant manuscript he undertakes to show that, when a cord is unwound off a cycloid, its end traces out a cycloid.[22] Given the cycloid FBA (Fig. 5.7) with axis FH and generating circle FLH, any tangential segment such as CB is parallel and equal to the chord FL. By Huygens's previous proof on the length of a cycloidal arc, FB equals twice the chord *FL*, and thus the line *DB* that equals the straightened *FB* is equal to 2*CB*. Is the curve described by *D* a cycloid? By still another property of the cycloid, Huygens shows that the arc *FL* of the circle *FLH* equals the line *LB*, which is itself equal to the line *FC*, because *FLBC* is a parallelogram. Drawing in the circles *MDC* and *CBG*, which are the same size as *FLH*, Huygens then claims the equality of arcs *FL*, *CB*, and *CD*, a fact that follows from the equality of their chords, and thus arc *CD* equals line *FC*. Therefore, *D* does indeed lie on a cycloid, because arc *CD* = *FC* is the mathematical statement of the mechanical definition of that curve.[23] In this manner, Huygens shows that the curve *FDE* described by the unwinding of a cycloid *FBA* is another cycloid.

Two important insights were necessary for Huygens's discovery that the normals to a cycloid define another cycloid. The first, which followed from his generalization of Wren's result, was the recognition that a curve with arc length twice that of a circle's chord is a semicycloid; the second, which was a direct mathematical representation of the mechanical foundations of the problem at hand, was the realization that the line *DB* equals the arc length *FB*. Although it arose from his research on clocks and thus had mechanical roots, the technique of unwinding a cord off a curve in order to find its length was completely general and could be applied to many curves other than the cycloid (see Chapter 7).

In the case of the cycloid and the application of its isochronism to a pendulum clock, the rectification of the entire arc *FBA* (Fig. 5.7) determines the length of the pendulum, because *FBA* represents one of the clock's curved plates, *FDE* the isochronous path of the bob, and *EA* the pendulum's cord, momentarily unrestricted at the midpoint of its swing. Since the "chord" on the generating circle that corresponds to point *A* is the diameter *FH*, the general rectification formula for the cycloid yields *FBA* = 2*FH*. Therefore, the length of the pendulum should be twice the diameter of the generating circle of the cycloid.[24] However, the length of a pendulum determines its time of swing, and consequently if a clock is to beat at a predetermined rate, the length ought to be calculated first and the cycloidal plates adjusted accordingly.

The steps to take in order to create a pendulum clock that, theoretically, beats isochronously are clear: Determine the length of the pendulum from the formula relating length and time, which in modern representation is $L = t^2 g/4\pi^2$ and which for Huygens was 9½ inches for a time of 1 second; bend two metal plates to conform to a cycloid that has a generating circle of diameter one-half that length; and hang the pendulum between them. As the pendulum swings against the plates, the bob will be altered from its circular path onto a cycloidal path equal in size to that which determined the plates. But a weight, such as the bob, falling along a cycloidal path will reach the bottom of the path in the same amount of time, no matter where in the path it begins its swing. Regardless of the amplitude of swing, therefore, the pendulum will beat uniformly.

With the requirements of an isochronous pendulum clock thus derived, Huygens turned to adapting his current clocks to the new design, and a flurry of activity continued into the new year. He wrote out instructions for his clockmaker on how to draw and form cycloidal plates and how to determine the length of the pendulum from the number of oscillations it should complete per hour, a figure dependent on the combination of gears chosen. He also calculated the deviations of the cycloid from the vertical, probably as a check on the accuracy of the bending of the plates. In addition, he began a table, which he maintained during the ensuing year, listing the variance between the solar day and the clock's time. His purpose was not to test the accuracy of the clock, but to chart the changing length of the day as the sun moved through its seasonal rounds – a necessary task if astronomers were to use his clock with precision.[25]

Huygens also announced his accomplishment to select friends, writing as early as December 6, 1659 – a mere week after his initial discovery of the isochronism of the cycloid – to his former teacher, Frans van Schooten:

I have been occupied with a new discovery that I deduced in these last few days in order to make my clock even more exact than it was hitherto. You know, I assume, that I added to these machines a pair of curved thin plates..., between which the suspended pendulum was swung; and because of that addition it performed in such a way that all swings of the pendulum, which otherwise were not totally *isochronous,* as I indicated in my book [the *Horologium* of 1658], were carried out in equal time. What, however, I never had expected I would discover, I have now at last hit upon, the undoubtedly true shape of the curves...that assures that all oscillations are very accurately made equal. I determined it by geometric reasoning, and I shall teach the artisans themselves how they may draw the curve with no difficulty. I venture to say that the discovery will not displease the very discriminating van Heuraet; for to me, certainly, it appears to be the most felicitous of all the discoveries that I have ever hit upon.[26]

The myth of continuity between the experimental determination of the plates in 1657 and the mathematical derivation of 1659 probably stems from this letter, and others like it, for it can be read as a justification of the claim that Huygens deliberately sought the isochronous curve because of his development of curved plates that caused approximate isochronous motion. Only a return to the manuscripts reveals the more convoluted path his discoveries followed.

Huygens's pride in his discoveries is understandable. In six short weeks he had moved from a search for the constant of gravitational acceleration to the derivation of the isochronism of the cycloid and the application of that knowledge to the pendulum clock by means of the cycloid's evolute.

THE EVOLUTE OF THE PARABOLA

What about the application of his new technique to that other isochronous path, the paraboloid? The reverse side of the very manuscript on which Huygens proves that the path (involute) described by unwinding a cord off of a cycloid is another cycloid contains a partial answer, for it shows his derivation of the evolute for a parabola (Fig. 5.8).[27] Having drawn the parabola HAB, its normals BEF and ADF, and auxiliary lines as shown in Figure 5.9, Huygens begins his search for the point of intersection F, which can be assumed to lie on the evolute MF when A and B are infinitesimally close, with a proportion similar to that used with the cycloid, namely $AF/FD = AC/DE$, obtained from the similarity of triangles ACF and DEF. Now, in a parabola all subnormals, such as EG and DN, are equal, because each is one-half the *latus rectum*, and therefore the two equations $EN = DE + DN$ and $EN = EG + GN$ yield $DE = GN$.[28] Because GN equals AO, the initial proportion becomes $AF/FD = AC/AO$, which in turn equals EK/KG because BE, BG, and BK cut proportional segments of AC and EK. Thus, $AF/FD = EK/KG$. Subtracting 1 from each side of this equation gives $(AF - FD)/FD = (EK - KG)/KG$, which becomes $AD/FD = EG/KG$. Because A and B are infinitesimally close on the parabola,

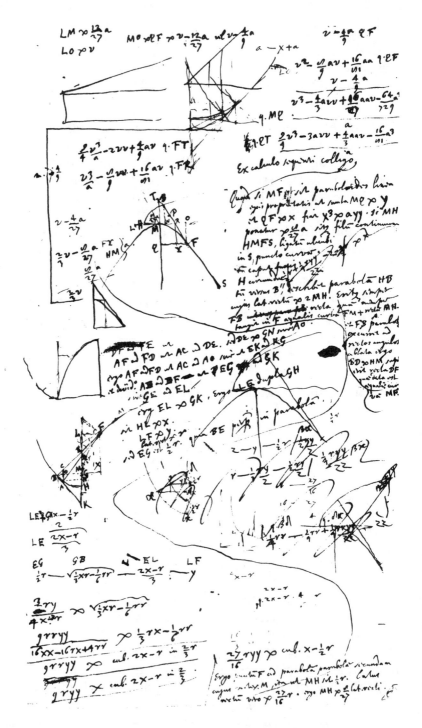

Figure 5.8. *Codex Hugeniorum*, f. 75r.

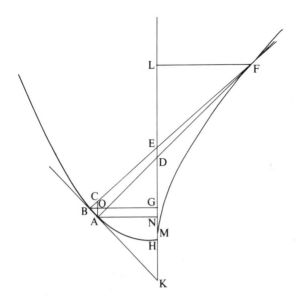

Figure 5.9. Deriving the evolute of the parabola.

AD/FD can be considered equivalent to *BE/EF*, which is *EG/LE* by the similarity of triangles *FLE* and *BGE*.[29] Thus, *AD/FD* is equal to both *EG/KG* and *EG/LE*, whence the subtangent *KG* of the parabolic involute equals the subtangent *LE* of the evolute being sought.

From this geometric conclusion Huygens proceeds algebraically to find the equation of the evolute. In the parabola *HAB*, *KG* equals 2*GH*; hence, *LE* (which equals *KG*) equals 2*GH*.[30] Now, *LE* + *GH* = *HL* − *EG*, and thus, when ½*LE* is substituted for *GH*, ³⁄₂*LE* = *HL* − *EG*. If *HL* is set equal to *x*, *LF* equal to *y*, and *EG* (as all subnormals) equal to ½*r*, where *r* is the *latus rectum* of the parabola, then ³⁄₂*LE* = *x* − ½*r* and *LE* = ⅓(2*x*−*r*). By the similarity of triangles *FLE* and *BGE*, *EG/GB* = *LE/LF*, or algebraically, noting that in the parabola *GB*² equals *rGH* or ½*rLE* and thus equals ⅙*r*(2*x* − *r*),

$$\frac{\tfrac{1}{2}r}{\sqrt{\tfrac{1}{3}xr - \tfrac{1}{6}rr}} = \frac{\tfrac{1}{3}(2x - r)}{y},$$

which reduces to ²⁷⁄₁₆*ry*² = (*x* − ½*r*)³. Thus, Huygens shows, the evolute of a parabola is a semicubical parabola with a *latus rectum* equal to ²⁷⁄₁₆ times the *latus rectum* of the given parabola.[31]

In his work on centrifugal force Huygens had discovered that a body, such as a pendulum's bob, would transit all horizontal circles on the surface of a paraboloid in equal times, as long as the centrifugal force were kept equal to the weight by a proper adjustment of the speed of spin. In other words, a parabola, when revolved about its axis, sweeps out an isochronous surface. Since the bob of an unrestricted conical pendulum remains at a fixed distance (equal to the length of the cord) from its pivot point, it moves on the surface of a sphere, in the same way that the bob of a simple pendulum moves along the arc of a circle. Therefore, the movement of the conical pendulum must be modified so that its bob travels a paraboloid, just as the simple pendulum is modified by cycloidal plates so that its bob moves along a cycloidal arc. In demonstrating that the evolute of a parabola is a semicubical parabola, Huygens discovered the necessary shape for the plate that would correspondingly adjust the conical pendulum. Theoretically, all he needed to do was rotate the pendulum and a plate shaped like a semicubical parabola about the axis; then as perturbations caused the cord to bank along the plate, its length would be altered so that the bob traced out a paraboloidal surface.

In this case, some minor differences are evident between the mechanical and the mathematical representations of the situation. Mathematically, the evolute of a three-dimensional paraboloid would be a three-dimensional semicubical paraboloid. Huygens could have designed a three-dimensional plate for his clock, but he chose instead to rotate a two-dimensional plate along with the pendulum. (Obviously, the two-dimensional plate really has three dimensions, but its width is discounted.) Of course, the surface swept out by the rotating plate is a semicubical paraboloid, and thus the mathematical result is the same, but the two-dimensional version allowed him to eliminate the friction on the rotating cord that would result if the pendulum had to rake along a solid surface. In addition, note that for the bob to travel along his given parabola *HAB* the cord must actually be slightly longer (by the distance *MH*) than the length of the unrolled plate. These are minor problems that show, nonetheless, that mathematizing the real world requires more than a mere one-to-one matching of points and positions.

Although the necessary tools – mathematical and mechanical – were readily at hand in 1659, apparently Huygens did not immediately apply his discovery of the evolute of the parabola to his horological work, for there is no evidence that he modified the conical clock to include the isochronous paraboloidal path until 1664.[32] This fact is rather surprising, since there is some evidence, albeit scant, that he did use the evolute of the cycloid to improve the simple pendulum and then to remeasure the constant of gravitational acceleration. Since he had originally used the conical pendulum to measure that constant, it seems strange that he did not improve the mechanism and remeasure. Of course, so little manuscript evidence remains of the experiments performed in 1659 that it is possible that he did contrive some sort of paraboloidal pendulum but that its existence passed unrecorded. It is more likely that the added dimension of motion for the pendulum was an unnecessary and mechanically unwieldy complication, given the alternative of modifying the (mathematically) two-dimensional simple pendulum.[33] All that can be said with certainty is that Huygens's attention in the waning days of 1659 focused on the mathematical, rather than the mechanical, consequences of his discovery.

THE GENERAL FORMULA

Almost immediately Huygens generalized the process by which he had found the evolutes of the cycloid and parabola. Although the original derivation is no longer extant, Huygens's calculations for the evolute of an ellipse, derived within two weeks of those for the cycloid and parabola, quite clearly depend upon a universally applicable formula.[34] Moreover, its application in the case of the ellipse leaves little doubt that Huygens's general solution in 1659 was essentially that published in 1673 as Proposition 11 of Part 3 of the *Horologium Oscillatorium.*[35]

The method, as described in the *Horologium Oscillatorium,* is a straightforward extension of his derivations for the cycloid and parabola. Given a general curve *ABF* (Fig. 5.10), any two normals such as *BMD* and *FNE* should also be tangent to the evolute at *D* and *E*. If *B* and *F* are indefinitely close, then both *D* and *E* can be approximated by the point of intersection *G* of the two normals. That is, *G* can be considered to be on the evolute, and Huygens's task is to define *G*'s position in terms of the given curve *ABF*. Because *F* and *B* are assumed infinitesimally close, the tangent to *ABF* at *B*, namely *BH*, can be considered to be the tangent at *F* as well, and thus the curve between *F* and *B* can be treated as a line segment. Draw *FPL* and *BK* perpendicular to the axis *AL*, and draw

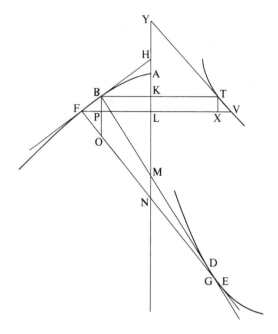

Figure 5.10. Deriving the general formula $BG/MG = (HN/HL)(KL/MN)$.

BPO perpendicular to *FPL*. By the similarity of triangles *BOG* and *MNG*, $BG/MG = BO/MN$. For the cycloid and the parabola this basic proportion was sufficiently amenable to manipulation to yield the respective evolutes directly. In general, however, Huygens found that the proportion was more manageable, and more revealing, if broken down into the composition

$$\frac{BG}{MG} = \frac{BO}{BP} \cdot \frac{BP}{MN}.$$

Since $BP = KL$ and $BO/BP = HN/HL$, the proportion can be rewritten as

$$\frac{BG}{MG} = \frac{HN}{HL} \cdot \frac{KL}{MN}.$$

This composed proportion becomes Huygens's fundamental formula for deriving any point *G* on the evolute of a curve *ABF*.

The two ratios on the right side of the equation are dependent only on the characteristics of the curve *ABF* and can be readily transferred into modern notation. The first ratio is a straightforward application of the first derivative, since HN/HL is equal to $(ds/dx)^2$ by the similarity of triangle FNL, triangle HNF, and the infinitesimal triangle having FB as its hypotenuse.[36] The second ratio involves the second derivative as well as the first, a fact seen by writing MN/KL as $1+(LN-KM)/KL$ and noting that the ratio $(LN-KM)/KL$ compares the change in the subnormals $(y\,dy/dx)$ to the change in the independent variable (x). Thus, in modern notation,

$$\frac{MN}{KL} = 1 + \frac{d(y\,dy/dx)}{dx} = 1 + \left(\frac{dy}{dx}\right)^2 + y\left(\frac{d^2y}{dx^2}\right)$$

$$= \left(\frac{ds}{dx}\right)^2 + y\left(\frac{d^2y}{dx^2}\right),$$

and KL/MN is the inverse of this value at any point. Combining the two factors gives

$$\frac{BG}{MG} = \left(\frac{ds}{dx}\right)^2 \frac{1}{(ds/dx)^2 + y(d^2y/dx^2)}.$$

Since $BM = y\,ds/dx$, the equation reduces to

$$BG = \left|\frac{(ds/dx)^3}{d^2y/dx^2}\right|,$$

which is the modern formula for the radius of curvature of a twice-differentiable curve.[37]

Huygens's geometric equivalent is, to repeat, $BG/MG = HN/HL \cdot KL/MN$, from which he could determine G's location on the normal BMG drawn to the curve at point B. Thus, the position of each point G on the evolute is defined by the characteristics of the original curve (the involute) infinitely close to the point B, which corresponds to the end of the unwinding cord in Huygens's physical representation of the process.

Obviously, in stating his formula for the pointwise derivation of an evolute, Huygens did not posit the modern restriction of twice-differentiability. However, he does go to some lengths in the *Horologium Oscillatorium* to justify the second ratio, KL/MN, after stating that the first, HN/HL, follows readily from the curve. Of course, the latter claim is fallacious if piecewise linear curves (which are not differentiable at their vertices) or other nondifferentiable curves are included, but for the small class of curves handled by Huygens it is an excusable premise.[38]

As to the ratio involving the second derivative, Huygens creates an auxiliary curve TV from the subnormals to the original curve (such as LN and KM) using the method of tangents, which he argues he can always apply. Further, he claims, because this new curve is geometric, its tangent (derivative) can always be found, and thus the ratio $KL/(LN - KM)$ $[dx/d(y\,dy/dx)]$ can be determined, from which he can then retrieve his ratio KL/MN, as has just been shown.[39] Again, he errs in assuming that the tangent to the new curve can always be found. However, as with the first derivation, for the curves available to him he has a valid method, one that yields some interesting results. In the case of the conic sections, he notes, the ancillary curve is always a line, and for the semicubical parabola the auxiliary curve is a cubical parabola.[40]

Fortunately, Huygens's first attempts at deriving evolutes in 1659 involved two very amenable curves. In the case of the cycloid, the decomposition of the initial ratio BG/MG is not even necessary, since the ratio is a constant 2. In the instance of the parabola, the second ratio KL/MN in the decomposition is constantly 1.[41] His subsequent attempt with an ellipse reveals how messy the calculations could become. Starting with an ellipse that can be represented algebraically as $4ay^2 + 4bx^2 = a^2b$, Huygens shows that the second ratio in the composed proportion defining the evolute is constant.[42] He determines the first ratio HN/HL by computing instead its equivalent HM/HK, which makes the composed proportion $(HN/HL) \cdot (KL/MN)$ an awkward

$$\frac{a^3 - 4ax^2 + 4bx^2}{a^3 - 4ax^2 - a^2b + 4bx^2}.$$ [43]

This algebraic expression for the ratio BG/MG constitutes the formula by which Huygens could determine the point G on the evolute with respect to its corresponding point B on the ellipse. Using this defining ratio, Huygens proceeds to find the general algebraic equation of the evolute itself with respect to a fixed origin, namely the point where the evolute meets the axis. Errors made in the lengthy series of calculations invalidate his resulting equation, although the process itself is perfectly legitimate. His second attempt, with the hyperbola $a^2b = 4bx^2 - 4ay^2$, even with a reduction to the equilateral case of $a = b$, requires a still longer set of algebraic derivations, but with no intervening mistakes at least his answer is correct.[44] Both evolutes turn out to be sixth-degree curves.

In addition to the awkwardness of the actual calculations, another difficulty with his method is the nonreversibility of the technique. In other words, although Huygens could find the evolute of a given curve, he could not find the curve's involute. Therefore, of the three initial curves whose physical properties prompted his theory, only two received companion curves, the evolute of the cycloid being a cycloid and the evolute of the parabola being a semicubical parabola. As for the third curve, the circular rim of the turning platform in *De Vi Centrifuga* was already the evolute; the curve in need of derivation was its involute, which described the path the released ball would take with respect to the man's position on the platform. Although Huygens could mechanically trace this involute by unwinding a thread off a circle, he had no means of constructing it mathematically.

The now familiar pattern of research can be seen repeating itself in Huygens's work on evolutes, in which he moved rapidly from a limited problem founded in mechanical concepts to the exploration of its theoretical consequences. Thus, after finding evolutes for two of the three curves that arose in his physical studies, he immediately applied his technique to the other two conics, neither of which had mechanical significance. Seven years earlier, this penchant for completeness had prompted Huygens to extend Archimedes' technique for finding the quadrature of the parabola to the ellipse and hyperbola in a treatise that won him much early recognition.[45]

Along with the generalization of his procedure for finding the defining proportion came the formalization of his discoveries into the beginnings of still another treatise. In the midst of the many pages devoted to calculating evolutes for the ellipse and hyperbola, preliminary theorems already appear.[46] One proves formally the special relationship of the involute to the evolute, namely that the tangents formed by unwinding a thread off a given curve are normals to the curve traced by the end point of the thread. The other, the purpose of which is not apparent at first glance, concerns parallel curves, showing that, if a second curve is equidistant from a given curve, then a normal to the first is normal to the second.

Having completed the derivation of the evolute of the hyperbola, Huygens returned to formalizing the foundations of his mathematical technique.[47] Paraphrased loosely, his short set of demonstrations combine to prove that parallel curves do not meet. Like the earlier, similar proposition, these theorems on parallel curves are not very meaningful in their own right. However, a comparison with the *Horologium Oscillatorium* reveals that the same propositions begin Part 3, in which Huygens presents the theory of evolutes, and culminate in a theorem that proves the uniqueness of the involute for any point of intersection on the evolute, thereby giving purpose to the previous demonstrations.[48]

In the uniqueness theorem, Huygens argues that by its very definition a second, distinct involute would be parallel to any given involute, because both would be perpendicular to the tangents of the evolute. Thus, only one of those involutes can meet the evolute at any specified point, because two involutes passing through the same point would imply that parallel curves intersect, violating the previous propositions. Huygens continues Part 3 of the *Horologium Oscillatorium* with propositions deriving the evolutes of the cycloid and parabola and then the general evolute theorem – all theorems founded on the discoveries of December 1659.

Having completed those studies, Huygens realized that he possessed the material for a new treatise, for soon thereafter he entered in his notebook an outline of a proposed second edition of the 1658 *Horologium*.[49] Greatly extending the scope of the original work beyond the description of a clock's design, the new edition would encompass all that Huygens had discovered in the previous two months, as well as supporting propositions. The list of items to be discussed includes: the use of a pendulum to determine longitude; the use of the pendulum as a universal measure; demonstrations on accelerated motion founded on Galileo's work; a report of Huygens's own experiments to measure fall and their agreement with Riccioli's result; the "great theorem" (*theorema magnum*), namely the isochronism of the cycloid; evolutes; and the astronomical application of the cycloidal clock.

As the new year began, Huygens's activities were geared toward completing the tasks necessary for this second edition. He had already revised his proof of the isochronism of the cycloid, labeling his new approach of December 15, which is very similar to the proof published in 1673, a "better demonstration."[50] This new proof treats the cycloid as an infinite collection of its tangents, thus returning to the method by which Huygens had unsuccessfully attempted to analyze circular fall in his earlier treatise on inclined planes. The "better demonstration" requires the results of the Galilean study, which explains the inclusion of the propositions of the latter as a prelude to the former in Part 2 of the *Horologium Oscillatorium*. Therefore, although Huygens did not succeed in applying the method pursued in the Galilean treatise to his original problem, he did prefer its approach when creating a finished, synthetic proof of his discovery. Not forgetting the theoretical extension of his work on evolutes, Huygens also determined evolutes for parabolas of higher order, and the results, like the other material, appear in the completed *Horologium Oscillatorium*.[51]

Indeed, the only portion of the *Horologium Oscillatorium* not substantially developed by the spring of 1660 was Part 4, which comprises propositions concerning the compound pendulum. Although Huygens attempted to deal with a weighted cord in 1659, he did not thoroughly study the problem and systematize his results until 1664.[52] However, the other four parts of the *Horologium Oscillatorium* were the theoretical consequences of his pursuit of the constant of gravitational acceleration begun in October 1659: Changing the problem to one involving centrifugal force, he developed *De Vi Centrifuga* and the theorems of Part 5. Mathematically analyzing the problem by comparing free fall with curvilinear fall, he developed the Galilean treatise and the isochronism of the cycloid presented in Part 2. Devising a mathematical technique to implement isochronism in his clocks, he derived the evolutes of Part 3. Finally, creating clocks that modeled his theories, he developed the precursors of those mechanisms described in Parts 1 and 5.

At each stage in this process of discovery, Huygens had addressed a single, rather specific question. The restrictive aspect of asking a narrowly focused question was offset by his thoroughness in pursuing the theoretical consequences of a problem, an example being his return to the mathematical formulation of Mersenne's experiment. One noticeable feature of this thoroughness is Huygens's need to summarize his results at each stage into a neat collection of theorems, a veritable small treatise on the subject at hand, the most famous being the posthumously published *De Vi Centrifuga*. The miniature treatises produced in 1659 during his work on the constant of gravitational acceleration were transferred almost wholly into the *Horologium Oscillatorium* of 1673 and thus are valuable aids in retracing the formation of this great work.

The *Horologium Oscillatorium* is above all a commentary on the effect of gravity on the pendulum regulator of a clock, and as such it is more a treatise on accelerated motion than a treatise on clocks per se. Certainly, in this masterpiece Huygens achieved the goal that he had proclaimed in his youth: to explain accelerated motion in a better way. It is in this light that Huygens viewed his work of 1659, writing to his friend, the astronomer Ismael Boulliau,

It is the principal fruit one could ever desire concerning the science of accelerated motion, which Galileo had the honor of first treating.[53]

6

Curvature

In solving his physical question, Huygens had developed a sophisticated mathematical tool that rightly takes its place in the *Horologium Oscillatorium* as one of his great discoveries. But just as his physics went beyond the narrow problem of the constant of gravitational acceleration to a broader discussion of accelerated motion, so his mathematics moved beyond the shape of curved plates to an analysis of one of the fundamental questions of seventeenth-century geometry. What, indeed, was the significance of the theory of evolutes as mathematics?

The evolute as a purely mathematical concept has long been relegated to a role in the history of curvature, a role determined by the very definition of curvature.[1] How, indeed, is the bending of a given curve to be measured? It seems obvious that a circle bends uniformly (that is, its curvature is constant) and that a small circle bends more sharply than a large one (that is, its curvature is greater) and, consequently, that the "curvature" of a circle might best be defined as the reciprocal of its radius. For any other curve the bending varies from point to point. Given a specific point, however, the most evident way to measure the bending is to assign to the curve the curvature of the circle that best approximates the curve in the immediate area of the specified point. The best approximating circle (labeled the "osculating circle" by Leibniz) can be derived by drawing normals to the curve at the given point and at a point infinitesimally near to it. The intersection of the two normals is the center of the approximating circle, the distance from that center to the given point on the curve is its radius, and from these relationships involving the approximating circle arise the names "center of curvature" and "radius of curvature."[2] Finally, the "curvature" at any point of a given curve is the reciprocal of the radius of curvature at that point.

Because each point on the evolute of the given curve (the involute) is derived precisely the way the center of curvature is calculated, the evolute can be defined as the locus of the centers of curvature of the involute. It does seem reasonable, even natural, therefore, to assign evolutes a prominent position in the history of curvature and thus to shift the search for the roots of the idea of curvature to early discussions of the locus that defines the evolute.

In his *Die Lehre von den Kegelschnitten im Altertum,* H. G. Zeuthen states that evolutes follow readily from Apollonius' work on normals in Book 5 of the *Conics.*[3] Repeated by T. L. Heath in his English synopsis of the *Conics* and by Paul ver Eecke in his French translation of the Apollonian masterpiece, Zeuthen's explanation has become a traditional fixture in histories of mathematics.[4] The implication of this view is that Apollonius contributed to the development of the theory of evolutes and that somehow, in a manner never specified, the history of curvature leads from Apollonius through Huygens to Newton and Leibniz. Carl Boyer's summary exemplifies this traditional interpretation: "The concepts of radius of curvature and evolute had been adumbrated in the purely theoretical work on *Conics* of Apollonius, but only with Huygens' interest in horology did the concepts find a permanent place in mathematics."[5]

How indebted is Huygens to Apollonius for the theory of evolutes? Is the standard outline for the history of curvature correct? Because the traditional argument is so firmly ingrained in the literature and, more important, because the question of mathematical influences uncovers a great deal about what Huygens was doing, it is worth examining in detail Huygens's role in the history of curvature.

APOLLONIUS AND MINIMAL LINES

Book 5 of the *Conics* is devoted primarily to determining whether a line drawn from a point to a conic contains a minimal segment.[6] That is, if point P (Fig. 6.1) is given and line CP is drawn, is line CM the smallest line segment that can be drawn to the conic at point C? As Apollonius himself proves, any such minimal line must be normal to the conic at C. Thus, the problem reduces to finding the normals to the conic from P, because any other line drawn to the conic from P cannot contain a minimal segment. Apollonius

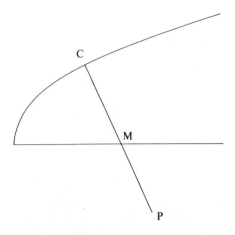

Figure 6.1. Apollonius' minimal line.

subdivides his problem into cases that are dependent upon the position of the given point from which the normals are to be drawn. After deriving the normals corresponding to points on the axis of a given conic, he considers the case in which the given point lies to one side of the axis and the normal is to be drawn to that portion of the conic that lies on the opposite side of the axis (as shown in the figure). Depending upon the position of the given point, there are none, one, or two normals. Modern commentators, taking their lead from Zeuthen, note that when only one normal exists the given point P is the center of curvature of the conic at the point C where the conic is intercepted, and in this manner Apollonius has presaged the concepts of curvature and evolute.[7]

Unfortunately, unlike modern commentators, the only knowledge that the mathematicians of the seventeenth century had of the contents of Book 5 of the *Conics* was a pair of brief remarks made by Pappus in his *Collection*. Listing ancient Greek mathematical works and their contents, Pappus notes that Book 5 of the *Conics* is about maximal and minimal lines, and in another passage referring to a proposition in Book 5, he accuses Apollonius of having violated the standards of mathematical verification by using a hyperbola in a situation where conics should not be introduced.[8]

In a curious twist of circumstances, direct knowledge of Book 5 was tantalizingly near, yet totally unavailable, to Huygens. First, a complete Arabic manuscript of Book 5, as well as Books 6 and 7, had been recovered at the beginning of the seventeenth century by an acquaintance of Huygens, but it was not translated nor even widely known until a century later. Jakob Gool (Golius), the other professor at Leiden (besides van Schooten) with whom Huygens studied and later maintained contact, had discovered the manuscript while traveling in the Near East in the service of Huygens's uncle. However, despite the fact that he was respected both as an Arabist and as a mathematician, Gool never was able to effect a translation, and his manuscript remained useless to everyone, including Huygens, until Edmond Halley translated and published it in 1710, along with his new translation from the Greek of the familiar first four books.[9]

Huygens was also aware of another Arabic manuscript of the last books of the *Conics* that had been discovered in the library of Leopold de' Medici at the beginning of the century and, like Gool's manuscript, unread for lack of translation. After many attempts at coordinating linguistic and mathematical interpreters, a very free synopsis was published under Giovanni Borelli's supervision in 1661, two years after Huygens's preliminary work on evolutes.[10] Thus, although Huygens was more aware than most of his contemporaries of the state of the Apollonian corpus, he had no greater knowledge of the content of Book 5.

Deprived of the original Greek texts, many mathematicians of the seventeenth century attempted "reconstructions" of ancient sources based on what few remarks and fragments remained. For example, in 1619, using Pappus' vague clues, Alexander Anderson published a reconstruction of what he considered to have been the major theorem of Book 5: to find the normal to a parabola from a given point not on the conic. His classically styled proof uses a hyperbola to determine the points on the parabola that are a minimal distance from the given point, thus incorporating the faulty technique criticized by Pappus.[11] However, neither his statement nor his proof of the problem is in fact similar to Apollonius' original study, which is much more involved and includes all three conics.

In 1653, Huygens's attention was drawn to Anderson's derivation of the normal to the parabola when a visiting friend questioned the validity of Pappus' objection to Apollonius' proof (reported by Anderson in his reconstruction). In support of Pappus, Huygens easily constructed a proof in which the offending hyperbola is replaced by a circle, the presence of which does not violate the ancient restrictions on mixing classes of curves.[12] It is clear from derivations recorded in Huygens's workbook that the conversation also turned to the problem of drawing normals to the cissoid and conchoid and prompted Huygens to apply the technique he used in his solution of the Apollonian problem to the determination of the inflection point of the conchoid, a result that he published the following year as Problem 8 of *Illustrium quorundam problematum constructiones*.[13] Thus, once again Huygens's research was instigated by a problem fortuitously posed by someone else, rather than by a systematic program initiated by himself.

Huygens describes the visit and his results in a letter to van Schooten, implicitly claiming himself a mathematician of merit by his ability to solve problems arising from ancient mathematics, just as he had previously gained recognition with his extension of the Archimedean quadrature of the parabola to the circle and the other two conics (*Theoremata de quadratura hyperboles, ellipsis et circuli, ex dato portionum gravitatis centro*, 1651) and as he would do again with his determination of pi (*De Circuli Magnitudine Inventa*, 1654).[14] Nothing seemed to spur Huygens's creativity more than the promise of proving he was a master of masters by solving, or at least extending, problems that his illustrious predecessors could not handle.

In the case of the normal to a parabola, since his derivation uses a circle rather than a hyperbola, Huygens's solution obviously has no similarity to Apollonius' approach. The circle that appears is not an osculating circle; it is not even centered at the given point from which a normal is to be drawn; rather it cuts the parabola at least once and as many as three times, where each point of intersection lies on a normal from the given point to the parabola. (It should be noted that Huygens's approach always yields one more normal than Apollonius' approach, because the former includes the one normal to the parabola that falls on the same side of the axis as the given point.)[15] Moreover, like Anderson and unlike Apollonius, Huygens does not distinguish the various cases into which the solution falls. Although he briefly notes that the number of normals can be 1, 2, or 3, he does not follow this statement with any attempt to determine the circumstances that yield a given number of solutions. In particular, the crucial case of only two normals (again, one more than in Apollonius' approach) is never discussed, and certainly no comments regarding centers of curvature or evolutes are in evidence.

Clearly, no link between Huygens and Apollonius was forged by either a direct or an indirect passage of ideas from Book 5 of the

Conics. Superficially, all the necessary ingredients were available; Huygens could have interested himself in the special case and, unlike Apollonius, he could have then invoked analytic geometry to determine the shape of the locus (a semicubical parabola), thereby deriving his first evolute. Why does the traditional scenario fail?

Nevertheless, I can truthfully say that I did not devote myself to it, believing to have already done more than was required,...so that I noted only the properties that presented themselves in the course of my proof, without straying to pursue others, uncertain of encountering anything that would pay me for my trouble.[16]

Huygens makes this disconcerting apology, perhaps never actually sent to anyone, while comparing the different methods that he, Johann Bernoulli, and Leibniz used in 1691 to discover the mathematical equation of the catenary. In the context of the situation, his explanation can be construed as no more than the specious excuse of one who had solved the immediate problem but had missed a powerful result noticed by the other two mathematicians.[17] Yet it can also be accepted as an honest statement of Huygens's attitude toward his scientific work. He had to have the goal in sight; he had to think the problem was significant and solvable. A nominal Calvinist he might be, but he did not espouse the Puritan work ethic.

However, if the standard history of curvature is correct, within Apollonius' work are the vestiges of curvature and evolutes, dormant until Huygens resuscitated them. In other words, the goal was – or should have been – in sight and the problem solvable. Quite obviously, Huygens did not see the question of drawing a normal to a parabola as a problem of curvature, either in 1653, when he addressed the Apollonian problem, or six years later, when he finally developed evolutes independent of any perceptible Apollonian influence.

The traditional history implicitly assumes that any mathematician who concerned himself with the special set of points now known to be the locus of the centers of curvature, which is automatically equated with the evolute, contributed to the history of curvature and that those contributions are cumulative, even when a historical figure makes no obvious use of the work of his supposed precursors. The mathematician need never ask how the bending of a curve should be described, and it does not seem to matter that he might have been asking some other, unrelated question for which that same set of points provides an answer. Yet Apollonius and Huygens were asking different questions. And it is on this point, not Huygens's undermotivated research habits, that the traditional scenario connecting Apollonius and Huygens fails.

A closer look at the *Conics* reveals that Apollonius' brilliant analysis has nothing to do with either the measurement of curvature or the derivation of an evolute. It is true that the geometric expressions of Book 5, when represented algebraically, reduce to the formula by which the center of curvature, and thus the evolute, can be found. Yet a fundamental element of any concept of curvature surely must be the idea that an intrinsic property of the curve is being measured, and nowhere in Book 5 is such an idea indicated. Apollonius does not systematically scan the locus of a conic and determine a corresponding center of curvature; rather, he surveys the field of external points and draws what normals he can. Even Zeuthen, upon whose account the traditional lineage rests, points out that in Book 5 Apollonius treats normals not as lines intrinsically related to points on the curve but as minimal line segments from points not on it.[18] To reiterate, Apollonius neither measured the bending of a conic nor derived a corresponding companion curve, and although the mathematical formulas might look the same, the theoretical and thus historical gap between his work on normals and later studies on curvature remains quite wide.

This denial might be countered by the question, if Apollonius was not concerned with curvature, what was his intent in deriving the point from which the normal is unique? Heath, in a footnote explaining the phrase "limits of possibility" (*diorismos*), provides the basis for an explanation. In his proof, Apollonius systematically moves along the axis of the conic and for each position on the axis (abscissa) proves that there is a position on the perpendicular to the axis (ordinate) such that, for any point on the ordinate closer to the axis, two normals to the opposite side of the conic can be found, and for any point on the ordinate further from the axis, no normal exists. If the chosen point is exactly at the demarcation position, one normal results. This point of uniqueness is important not because it says something about the given conic and its shape, but because it delimits the categories into which the solutions along the ordinate fall. Its derivation determines the "limits of possibility" both by showing that a solution is possible and by revealing "in how many ways it is possible."[19] Thus, a viable explanation of Book 5 can readily be constructed from traditional Greek mathematical precepts without recourse to modern concepts. Apollonius was solving a different problem, and only through hindsight do his theorems on normals become the vestiges of curvature and evolute.

In fact, startling though it may sound, Huygens's early work on evolutes also had nothing to do with curvature, and only hindsight turns his derivations of companion curves for the cycloid and conics into a method for determining curvature. Although the geometric equivalent of the radius of curvature appears in his derivations of evolutes in 1659, Huygens never sought to measure the bending of the involute. He, too, was asking a different question.

It is true that Huygens indirectly dealt with curvature in a very qualitative way. For example, in his theorem on parallel curves he specifies that the second curve be "outside" the first, "that is, its curvature [*cavitas*] depends upon the convexity" of the first, but he never investigated this dependency.[20] His definitions of evolute and involute require no more than that the two lines be "curved to the same side."[21] Nothing in his manuscripts indicates that he ever thought to compare the amount by which the curves bend, that is, to measure their curvature. Only the geometric relationship between the two mattered.

Another potential catalyst that failed to prompt an analysis of curvature was Huygens's study of the inflection point of the conchoid, begun in 1653 simultaneously with his work on the Apollonian problem. Certainly, the point at which a curve reverses its bending, changing concavity from one direction to the opposite, seems an appropriate stimulus for an investigation into the nature of curves, and in 1659, two months before his intensive investigation of evolutes, Huygens was busy refining his original study.[22] However, his derivation of the inflection point of the conchoid remained an isolated case, another successful treatment of a problem descendant from ancient mathematics that was never generalized to other inflecting curves or in any way developed beyond the immediate domain. Furthermore, aside from revealing his general sophistication in handling infinitesimals, nothing in the work appears to hold significance for his development of evolutes. Chronologically close, evolutes and inflection points stood apart in Huygens's work, neither generating a comprehensive study of either tangency or curvature.

Indeed, one compelling argument for denying that Huygens was interested in measuring the bending of a curve is the fact that he wrote no small treatise outlining definitions and theorems regarding curvature. Surely, if the concept had captured his attention, he would have followed his customary practice of generalizing his work and detailing its salient features. Significantly, the *Horologium Oscillatorium* itself contains no such analysis linking curvature with evolutes.

Like Apollonius, Huygens was not concerned with determining an intrinsic property of the given curve. His attention was focused wholly on the derived companion curve; the evolute, not the involute, interested him. Furthermore, his approach was holistic, emphasizing the entire curve and its equation, and although he derived the evolute pointwise, he eliminated the infinitesimal situation as soon as possible, a procedure he frequently followed in his derivations. This attitude allowed him to skirt the shaky methodological foundations of infinitesimal analysis, but it also encouraged him to ignore the instantaneous aspects in favor of the resulting totality.

NEWTON AND MEASURE

In comparison, Isaac Newton's early work on curvature, begun in December 1664, is explicit and thorough. Newton actively seeks the "quantity of crookednesse in lines" and compares the curve with an approximating circle. His procedure for finding that approximating circle begins in the same way that Huygens's derivation of the evolute begins; two normals are drawn to points infinitesimally close on the curve, and their intersection determines the desired point. Newton's derivation is much more Cartesian, however, since he sets the normals equal and applies Hudde's algorithm for multiple roots. Perhaps the Cartesian method of equal roots predisposed Newton to visualize the normals differently than Huygens did, although both men would have been well aware that true equality between normals held only in the limit. For whatever reason, only Newton clearly describes the point of intersection as the center of a circle whose radii are the "equal" normals and whose curvature is the same as that of the given curve.[23]

Unlike Apollonius and Huygens, Newton unambiguously directed his research toward discovering the inherent qualities of a given curve. For example, having determined the curvature of the various conics, Newton immediately continued on to the question of "greatest or least crookednes."[24] Moreover, his entire study of curvature was just one aspect of his preliminary work on the calculus, which also included theorems on normals and tangents. In May 1665 he wrote a rough treatise outlining some of these discoveries, and in October 1666 he produced a more polished manuscript detailing his method of fluxions, including an expanded analysis of curvature.[25] Thus, the aspects of curvature missing in Huygens's work are readily apparent in Newton's early fluxional analysis.

In addition, however, Newton's study of curvature led him to many of the results Huygens did discover, results regarding the arc length of the evolute that will be detailed in the next chapter. Already in May 1665 Newton noted that the curve formed by the centers of curvature of the parabola has as its tangents the normals of that parabola and is measured by them: "By this meanes ye length of as many crooked lines may bee found as is desired."[26] This remark is the foundation of a more thorough analysis of rectification that Newton presents in the 1666 treatise beginning with the problem: "To find such crooked lines whose lengths may bee found. & also to find theire lengths."[27] In this way, although he obviously did not use Huygens's terminology, as yet unpublished, Newton derived the evolute and knew that the distance along its tangent to the involute measured the arc length of the evolute. Thus, he duplicated Huygens's results in an entirely different context.

In 1671, continuing his independent path to the calculus and, coincidentally, to a study of rectification, Newton completed an expanded treatise on fluxions (*De Methodis Serierum et Fluxionum*) that combined the 1666 tract with still another work on the calculus (*De Analysi*).[28] When John Collins, self-appointed protector of British mathematical discoveries, urged Newton to publish the 1671 treatise because rumor had reached him that Huygens intended to bring out a book on "Dioptricks *et de Evolutione Curvarum*," Newton was not at all perturbed, Collins reported to a correspondent, "supposing it altogeather improbable that their Hypotheses or Deductions can be the same."[29] Of course, Newton could not have ascribed Huygens's technical meaning to the work *evolutione* and, consequently, probably understood the phrase to mean no more than "dioptrics and on the generation of curves." Thus, when the *Horologium Oscillatorium* finally appeared in 1673 with an extensive discussion of rectification, Newton must have been surprised by the similarity of some of their deductions after all. His letter to Oldenburg in which he thanks the intermediary for his copy of the *Horologium Oscillatorium* includes a polite claim of originality:

The rectifying curve lines by y^t way wch M. Hugens calls Evolution, I haue been sometimes considering also, and haue met wth a way of resolving it wch seems more ready and free from ye trouble of calculation than y^t of M. Hugens. If he please, I will send it him. The Problem also is capable of being improved by being propounded thus more generally.[30]

Newton's claim is astonishingly modest, especially when measured against the vitriolic priority squabble with Leibniz that developed later in his life over the invention of the calculus. Although he argues that his method is better and more general, he seems quite willing to share it with Huygens. Perhaps early in his career he could be more deferential to an older scientist whose work he knew and whose opinions he respected, even when the two men differed as strenuously as they did on his 1672 paper on colors. In fact, the vast bulk of the letter to Oldenburg concerns a reply to Huygens's objections to the colors paper, so perhaps Newton was only trying tactfully to avoid another argument while the first was yet to be resolved. Nevertheless, the unanswerable question pricks the imagination: Was Newton, still relatively untouched by disputes, prepared to send Huygens a report of his unpublished treatise on the calculus?

Alas, Huygens never responded to Newton's overture when it was conveyed by Oldenburg, who even offered his assistance in the matter.[31] Probably Huygens felt that he had nothing to learn from the younger man, since Newton claimed only to have a different way to deal with problems already solved by Huygens. In a similar situation, Huygens's early response to Leibniz's calculus was lukewarm because he felt that he had an equivalent method for solving the problems at hand and he did not appreciate Leibniz's systematization of technique for its own sake. Presumably, Huygens would have likewise undervalued Newton's treatise, even though, as we now know and as he was eventually to realize, his own technique was not as powerful as either the Newtonian or the Leibnizian calculus.

Clearly, any communication between the two men in 1673 would have had no effect on either Huygens's work of 1659 or Newton's research of 1665–6, and this obvious conclusion applies not only to the untransmitted treatise of Newton but to the published *Horologium Oscillatorium* as well. Indeed, Newton's response to Collins is proof that he knew nothing of the theory of evolutes before his receipt of the *Horologium Oscillatorium*. Thus, although Huygens's analysis of evolutes antedates Newton's discussion of curvature, chronological priority of discovery does not make Huygens a precursor of Newton in the development of a theory of curvature or of rectification. True, their "deductions" regarding the rectification of curves are similar, but the "hypotheses" that guided their research were, as Newton predicted, quite different.

Once again the traditional history of curvature errs by trying to link two separate studies that only superficially deal with the same subject. Not only was Huygens not influenced by Apollonius, Newton was not influenced by Huygens. Nor, it should be added, was Gottfried Wilhelm Leibniz influenced by the man often described as his mentor, although in this case the potential link between the two mathematicians is more complicated and, hence, more difficult to analyze.

LEIBNIZ AND STRUCTURE

An exchange of letters between Leibniz and Huygens focuses on the question of influence and illuminates some of the difficulties of tracing the history of an idea like curvature. Many years after his initial work on evolutes, Huygens explicitly derived a radius of curvature and thereby instigated a minor priority squabble with Leibniz regarding the osculating circle.

In 1690, Jakob Bernoulli challenged mathematicians to find the true shape of the catenary, and as part of his response Huygens derived the catenary's equation, found its evolute, and calculated its arc length.[32] By replacing the catenary at its lowest point (the vertex) by a circle, he was able to eliminate from his formula for the arc length factors that depended on the number of sides in the polygon that he used to approximate the catenary. By this means he was able to reduce his equation to a simple relationship: The length of the catenary from the vertex to a given point equals the radius of the approximating circle at the vertex multiplied by the ordinate value of the given point and divided by the subtangent; in modern notation, $L = r\,dy/dx$, where r is the radius of curvature at the vertex. The result is quintessential Huygens in its elegance and simplicity, with all extraneous material cut away and only the geometric, almost self-evident truth left.

However, although he derived the radius of curvature at the vertex, Huygens applied it in a very strange way, using it merely as a unit of measure with which to determine arc length. Never seeking the radius of curvature at any other point, he did not interpret the evolute as the locus of the centers of curvature – a supposedly obvious connection. Of course, he did derive the evolute, and consequently he did indirectly determine the radius of curvature for any arbitrary point on the catenary. But the manuscripts show no evidence that he was consciously analyzing curvature even at this stage.

Huygens encoded his results in anagrams, sent them to Leibniz in October 1690, and revealed his solutions to both Leibniz and the readers of *Acta Eruditorum* in May 1691.[33] In the May explanation, Huygens introduced the term *radius curvitatis* as the "semidiameter of the greatest circle drawn through the vertex that falls completely inside the curve."[34] This appears to have been the first published definition of the radius of curvature. However, it is perhaps significant that Huygens defined it only at the vertex, for he might have intended it to function as a mathematical constant, a parallel to the *latus rectum* of the parabola, rather than a variable entity akin to a normal. In other words, although Huygens used the radius of curvature as an analytic description of an intrinsic property of the catenary, it is unclear whether he was anticipating the modern measure of an instantaneous event or was attempting to fit the catenary to ancient models.

In response to Huygens's paper, Leibniz tactfully brings up his own research regarding the osculating circle:

I did not consider the curve that by its evolution can produce the catenary. However, I see that it is useful to think about it in one's researches. I do not know, Sir, if you have noticed a small *discours de Angulo contactus et Osculi,* which I published in the *Acta Eruditorum* in June 1686. There I argue that the direction of a curve ought to be represented by its tangent, because that line always maintains the same direction: and the tangent only makes with the curve an angle of contact, which is less than any angle between lines. But the curvature [*courbure*] or flexion of the curve at each point ought to be represented by the circle that touches it there the most exactly, or which kisses it, for the circle always maintains the same curvature; and the circle that kisses only makes with the curve an *angulum osculi* as I call it, which is less than any angle of contact between circles. This agrees with what you say, Sir, regarding the radius of curvity. That is why one would do well to consider the latter when studying curves. And the centers of the circles measuring the curvature fall on your evolute. Perhaps, it would be good to continue the progression and to study what curve would be appropriate as the measure of the osculation of second degree.[35]

Huygens replies with a reference to Part 3 of the *Horologium Oscillatorium:*

You speak to me, with regard to the curvature of the catenary, of your *discours de angulo Contactus et Osculi.* You can well believe that in reading it I did not find this idea new, because those kinds of contact enter naturally in my Evolutions of Curved Lines.

In addition, since Leibniz had remarked in his paper that a circle with a diameter equal to the *latus rectum* of a parabola has the same curvature as the parabola at its vertex, Huygens, apparently feeling the need to squash any further question of originality, cites a letter he wrote to van Schooten in 1654 regarding revisions for the second edition of the latter's *Geometria* (1659). In that early letter he had briefly noted that at any point of a parabola, not just at its vertex, a circle can be found passing through that point and sharing the parabola's tangent at that point. Subtly expanding its application and appropriating Leibniz's language, Huygens tells Leibniz about the result sent to van Schooten, concluding, "As with the other conic sections, it is not only at the vertex that a circle can kiss the parabola; moreover this applies many times to other curves although it appears you have said nothing of it."[36] Undaunted, Leibniz rejoins:

I well believe that you have seen the circle that is described at a point of the Evolute, and of which the radius is the minimal line that one can draw from that point to the described curve; but perhaps you had not thought at first to consider it as the measure of the curvature; and I, when I had considered the maximum circle that touches the curve interiorly as the measure of curvature or of the angle of contact, I did not think of evolutions.[37]

All the frustrating difficulties of tracing an idea to its inception are displayed in this vignette. Although Leibniz's reference to the 1686 paper need not be interpreted as a veiled claim of priority regarding the osculating circle, Huygens appears to have read it as such, and certainly Leibniz substantiates the interpretation with the attempt in his rejoinder to partition the territory, granting Huygens priority on evolutes while retaining osculating circles for himself. Is Leibniz's division valid? And if Leibniz originated the osculating circle, does it follow that he influenced Huygens's use of it?

It is clear from Huygens's habits that his introduction of an approximating circle into the catenary problem really did derive from his own early work and not from Leibniz's paper. His citation of the letter to van Schooten supports the hypothesis that he was drawing a parallel between the parabola and the catenary and applying techniques that he learned regarding the former to the latter. Knowing that a parabola can be approximated by a circle, he could easily have been led to using a circle in place of the catenary, a curve he knew to be close to the parabola in shape.

Moreover, it is highly unlikely that Huygens originally read Leibniz's paper with much attention. His tendency to read cursorily is exemplified by his dealings with Nicolas Fatio de Duillier and Leibniz over the inverse tangent problem. At one point he admitted to Leibniz that he had not yet seriously studied solutions that Fatio had sent him a year before, even though he was supposed to be mediating their dispute.[38] Indeed, his initial lack of progress in assimilating the calculus was in great part due to his inattentiveness to Leibniz's explanations. Like many mathematicians, now as then, he preferred solving the problem himself, using his own notation and techniques.

Of course, it might more readily be expected that Huygens influenced Leibniz. Although descriptions vary regarding the extent to which Huygens acted as mentor to Leibniz during their encounters in Paris in the years 1673–6, Leibniz was very much the neophyte when Huygens presented him with a copy of the *Horologium Oscillatorium* in 1673. Thus, it can be presumed that Huygens's references to the book and to research done when Leibniz was only eight years old were intended not only to deny Leibniz a role in Huygens's work on the catenary but also to remind Leibniz to whom he was obliged for his introduction to contemporary mathematics. With his statement that osculating circles arise naturally from evolutes, Huygens seems to conclude that Leibniz is the one in debt.

In his design for the clock of 1657, Huygens had depicted the path that the bob should travel as a series of circular arcs through whose centers of rotation a plate must be curved. In 1659, knowing the exact path (involute) the bob should take, he was able to derive mathematically the shape of the plates (evolute) by using the intersection of the normals to two points infinitesimally close on the involute to determine a point on the evolute. In this infinitesimal setting, the two normals could be considered equal radii of a circle that is centered at the point of intersection and approximates the involute over the infinitesimal arc between the two points from which the normals are drawn – Newton's approach to curvature. Thus, osculating circles do arise naturally from evolutes. However, although the 1657 attempt clearly demonstrates that Huygens could consider the involute as a collection of approximating circular arcs centered along the evolute, in fact, his derivation of the basic proportion by similar triangles does not set the normals equal or appeal to circles in any way.[39] Additional proof that he was not thinking of approximating circles is the argument from silence, for he said nothing about such circles in the *Horologium Oscillatorium.*

Perhaps, while reading the *Horologium Oscillatorium,* Leibniz nonetheless perceived the vestiges of the approximating circles in Huygens's discussion. Such unconscious transmission of the germ of an idea is impossible to trace, since there is no evidence in Leibniz's published manuscripts of textual analysis leading to the new theory or of borrowed terminology.[40] To his credit, and against the conclusion that there was a conscious absorption, is Leibniz's habit of publicly acknowledging his debts to Huygens, if for no other reason than the reflected glory he seemed to feel accompanied their relationship. Moreover, like Huygens, Leibniz was tied to his own techniques and did not readily absorb whole the work of others.

Not only did Huygens and Leibniz develop their ideas regarding approximating circles independently; they put those ideas to such very different uses that it would be grossly misleading to label this a case of simultaneous discovery. Leibniz correctly identified that difference in his letter of rejoinder when he pointed out that he was concerned with measure and not with "evolutions." Moreover, in Huygens's work the substitution of a circle for the curve under consideration was an approximation technique and, as such, was a means and not an end.

When he voiced his *déjà vu* complaint to Leibniz, Huygens pointed to his own recognition that the parabola could be approximated by a circle, and it is true that he had achieved very fruitful results by interchanging the two curves. His first success using this approximation was his determination of the center of gravity of a circular section, published in *De Circuli Magnitudine Inventa;* his second was his derivation of the formula for centrifugal force; his third was his discovery that the cycloid is isochronous.[41] It is significant, however, that in each case Huygens approximated the circle by the parabola and not the other way around. A useful analytic tool, the parabola reduced complex relationships to manageable equations, and a great part of its success in two of the cases stemmed from the fact that it also represented geometrically the physics of fall. This use of the parabola could not readily be transferred to a general analysis of curvature, since each application was treated differently according to the mathematical needs.[42]

Even the case of the catenary, in which Huygens does use an approximating circle, is no exception, for when he makes the substitution in his original derivation Huygens notes that a parabola could just as easily be used as the circle.[43] More precisely, either one would give him an answer freed from the extraneous variable because the osculating parabola and osculating circle are readily interchangeable, with the *latus rectum* of the first being equal to the diameter of the second.

Further, because the approximation, whether by a parabola or by a circle, was only a tool, it had no independent life and required no theoretical justification. It was merely a construction device used to achieve the answer, and once the goal was reached the scaffolding was removed, leaving the result free standing – the paragon of mathematical aesthetics.

In contrast, Leibniz used the osculating circle as part of a general scheme with which to describe a curve: The tangent gave a first-degree approximation; the osculating circle, a second. His was a vision of infinite generality, of form underlying the particular. His method for deriving the circle demonstrates his unique approach. He does not begin with normals and determine the center of the circle from their intersection; rather, he looks at all circles that cut the curve in four points and takes the limiting circle where the points coalesce.[44] By the very nature of the derivation, the locus of the centers (the evolute) is deemphasized in favor of the approximating circles themselves. In Leibniz's approach, the scaffolding was the structure and could not be removed. The mathematical approximation was no longer a means to some specific solution but rather a defining characteristic of the curve. And it is Leibniz's vision of structure that whole generations of mathematicians have sought to validate and expand.

In summary, Apollonius, Huygens, Newton, and Leibniz had very different concerns at hand when each derived a mathematical expression that can be translated into the modern formula for the center of curvature. Apollonius was searching for the limits on the possible solutions to the problem of drawing minimal lines to a conic section; Huygens was deriving the companion curve that held a privileged geometric relationship with a given curve; Newton was measuring the quantity of crookedness of a curve; Leibniz was developing a general theory of curves. They did not influence one another in any significant sense, and their isolation and the differences among their intentions are severe enough to undermine any arrangement of their works into a chronological progression that is meant to function as an outline for a history of curvature.

When the interests of the four mathematicians are conflated, as in the traditional account, it becomes very difficult to assess accurately an individual's creative contribution. Obviously both Newton and Leibniz provided elements necessary for the development of the idea of curvature, with Newton concentrating on curvature as measure and Leibniz on curvature as structure. However, the man who introduced one of its basic terms (the radius of curvature) actually does not belong to the history of curvature, except as a cautionary example. Huygens's treatment of curves shows how the supposedly obvious required conscious development.

Once made conscious, however, the concept of curvature did appear self-evident. After his exchange with Leibniz over the catenary, Huygens never again derived a radius of curvature, with one exception. In July 1692 the Marquis de l'Hospital sent Huygens his rectification of the logarithmic curve (also called the logistic or exponential curve), which Huygens had been among the first to study. The new discovery prompted Huygens to reexamine the curve and to find, among other results, the point of greatest curvature (*ubi maxima ejus curvitas*). In his proof he follows the technique used in his general derivation of evolutes in order to get the curvature and then applies Hudde's maximization rule.[45] Thus, curvature truly is a natural extension of his work on evolutes, just as he had claimed to Leibniz.

There is another "natural" extension of evolutes, one for which Huygens has a much more legitimate claim to recognition. As usual, he was originally drawn to this application in part because of his habit of responding to outside stimuli. Newton identified the topic when he referred to the method of "rectifying curve lines" that Huygens discusses in the *Horologium Oscillatorium*. Indeed, whereas Huygens says nothing about the connection between evolutes and curvature in his book, he says a great deal about the measurement of the arc length of an evolute. Rectification, not curvature, was the mathematical application to which Huygens turned in 1659.

7

Rectification

Mathematicians of the late seventeenth century focused on two problems that derived from Greek geometry, particularly from attempts to determine the ratio of the circumference of a circle to its diameter, the value now called pi. The first, which was related to the question of measuring the circumference of a circle, was rectification: how to find a straight line (*recta*) equal in length to the arc of a curve. The second problem, which was tied to the question of ascertaining the area of a circle, concerned quadrature: how to find a square (*quadratum*) equal in area to a given two-dimensional space. Of course, the exact statement of these problems changed over the centuries, and thus the seventeenth century inherited a curious conglomerate of specific questions within the two broad categories.

However, the standards by which solutions to questions of rectification and quadrature were judged successful remained remarkably constant in the centuries separating Greek and seventeenth-century mathematicians. With the formalization of analytic procedures in more recent centuries, questions involving area and length are now usually solved by trivial integrations that yield a numerical answer. But in earlier times the solutions had to be geometrically constructable, and even in the seventeenth century algebraic solutions had to be reducible to geometry by means of standard constructions like those outlined by Descartes in his *Géométrie* and codified by van Schooten in his Latin translation and commentary, *Geometria*.[1]

Because Archimedes had already found the area under the parabola by geometric methods, the feasibility of performing at least some quadratures was never doubted.[2] On the other hand, although methods for finding the length of the circumference of the circle were proposed in ancient times, they all foundered on the requirement of constructability.[3] Not unnaturally, therefore, mathematicians came to believe that geometric rectification was impossible, and philosophers obligingly set forth strictures proscribing comparison of curved and straight. Thus, Descartes was following a tradition going back to Aristotle when he wrote in the *Géométrie*, "The ratios between straight and curved lines are unknown, and even, I believe unknowable to men, so that we cannot thereby reach any exact and assured conclusions."[4]

Within two years of the publication of the *Géométrie* Descartes himself provided evidence that his claim was erroneous, when, in a letter to Mersenne, he commented on the linear relationship between the logarithmic spiral and its radial arm. Unfortunately, his remark was not widely known until his letter was published posthumously in 1657.[5] Likewise, attempts by Evangelista Torricelli, Gilles Personne de Roberval, and Pierre de Fermat to deal with rectification all passed relatively unnoticed, partly because these early rectifications were not completely adequate (they involved transcendental curves), partly because they were not communicated to the general scientific community (perhaps because they were realized to be inadequate).

However, in 1655 the problem of rectification gained notoriety when Thomas Hobbes launched a series of caustic attacks on English mathematicians, particularly John Wallis, who responded in kind. Scoffing that the claim of impossibility was merely an excuse for incompetence, Hobbes proceeded to rectify both the circle and the parabola as well as to square the circle. His "proofs" were fraught with errors, which were quickly exposed by the mathematicians but excused by Hobbes as incidental to the main argument.[6]

A more private, though sometimes no more polite, argument was occurring at the same time between Huygens and followers of Grégoire de Saint-Vincent. Four years earlier, in an appendix to his *Theoremata de quadratura,* Huygens had refuted Grégoire's claim to have squared the circle.[7] While the battle between Hobbes and Wallis raged, Huygens was still being harassed by Grégoire's defenders, who – like Hobbes – refused to accept rebuttal.[8]

In 1657 Huygens was to discover that the arc length of the parabola is directly related to the area under the hyperbola, thereby tying the problems of rectification and quadrature together. (Archimedes was actually the first to have made this association when he showed that the area of a circle is equal to the circumference times one-half the radius.)[9] Historians of rectification have presumed a causal connection between Hobbes's attempts to rectify the parabola and Huygens's later work:

Hobbes stubbornly persisted in believing that his rectifications were legitimate. Huygens was among those who tried in vain to put Hobbes straight; and in this connection Huygens in 1657 found that the rectification of the parabola could be reduced to the quadrature of the hyperbola.[10]

In fact, Hobbes seems to have had no influence on Huygens's work. Because of the controversy over Grégoire de Saint-Vincent's erroneous quadrature, Hobbes apparently desired Huygens's approbation of his own mathematical efforts. However, Huygens did not respond to the solicitation made by a mutual friend. Indeed, the only extant evidence of Huygens's response to the British antagonists is a note to Wallis saying that he had spotted the mistake in Hobbes's rectification of the parabola, which Wallis had criticized in a previous letter, and expressing surprise that Wallis would give credibility to Hobbes by replying so voluminously in print.[11] Although his persistent desire to stay at the top of any area in which he had previously invested his reputation might have fueled his ambition to participate in the contest, the acrimony of the debate seems to have kept him aloof, at least for the time.

Moreover, Huygens recorded his discovery concerning the parabola's relationship to the hyperbola well over a year later, which is too long a span of time to justify the claim of influence without some kind of substantiating evidence. Huygens's reaction to a stimulus was usually evident within days of his exposure, even in cases, like the Pascal challenge problems, where his consequent discoveries were not all immediately forthcoming. Hobbes might have posed the problem directly – find the arc length of the parabola – but the more immediate influence on Huygens seems to have been a different, less direct stimulus, one that will be discussed later in this chapter. Finally, even in England, Hobbes's nettling did not provoke a flurry of activity by the leading mathematicians.[12]

However, the British mathematicians soon learned that Hobbes was correct on at least one point. Geometric rectification was possible. Wallis had already argued that analytic rectification could be achieved by means of an infinite series involving infinitesimals of arc length (in modern garb, $ds = \sqrt{dy^2 + dx^2}$). In the summer of 1657, adapting Wallis's approach to geometric methods, William Neil was able to find a straight line equal to the arc length of a geometrically representable curve, namely the semicubical parabola.[13] Neil did not publish his geometric rectification but did communicate it verbally to his interested friends, including Wallis, Wren, and Brouncker, each of whom in turn tried to prove the result himself. One year after Neil rectified the semicubical parabola, Wren rectified the cycloid. Again the solution was not published at once, but in this case the result did become known beyond the immediate circle of British mathematicians when Pascal announced Wren's achievement in his *Histoire de la roulette* of 1658.

Pascal's announcement, even more than Hobbes's pontifications, brought the problem of rectification to the fore, because for the first time the general mathematical community knew that an answer was attainable. Like Archimedes' quadrature of the parabola, Wren's rectification (as well as Neil's comparatively unknown one) was a promise of feasibility. It did not guarantee that every curve could be geometrically rectified and squared, but it did mean that attempts to solve the two problems might be successful, might "pay" the researcher for his effort.[14] No wonder that Huygens praised Wren's discovery, and no wonder, too, that he tried to generalize the result to an arbitrary section of the cycloid, a success described in Chapter 5.

By late 1659, when Huygens was undertaking his research on evolutes, the problem of rectification had become so important that Wallis published, as an open letter to Huygens appended to his alternative history of the cycloid (*Tractatus Duo*), the proofs of both Neil and Wren, along with a short history of rectification in England.[15] As it so often was, Wallis's motive was to protect the priority of his countrymen against usurpation from abroad. He had reason to worry, because a few months earlier Huygens had informed him that the new edition of van Schooten's *Geometria* would contain a general rectification procedure developed by Hendrik van Heuraet and that, in particular, van Heuraet could find the arc length of the semicubical parabola.[16]

PRIORITY AND THE PARABOLA

Huygens was scarcely a disinterested reporter of van Heuraet's triumph, for quite clearly the purpose of his letter was to draw attention to his own contribution to rectification. In the fall of 1657, Huygens had made two related discoveries regarding the parabola.

The first was his aforementioned proof that the arc length of a parabola could be directly related to an area under the equilateral hyperbola. The second was a proof that the area of the surface swept out by a parabola revolved about its axis was directly proportional to the area of the circular base of that paraboloid.[17]

Huygens had immediately hinted about his discoveries to his two current mathematical correspondents, van Schooten and René François de Sluse, telling Sluse:

Moreover, I love those problems in which an extraordinary discovery is a truly easy calculation. A few days ago [I found] two new ones concerning the parabola (or so they certainly appear to me and I have mastered this excellent discovery), for which I am currently writing down the entire study.[18]

Two months later he informed both correspondents of the second result, stating that he wanted to withhold the first because he was having some difficulty (unspecified) with it.[19] Perhaps he was dissatisfied with having reduced the rectification of the parabola only to a quadrature rather than to an exact value, although that achievement alone would have been applauded. Perhaps, as his editors suggest, he was displeased with his casual derivations, although both discoveries were given formal proofs in the "entire study" cited by Huygens to Sluse.[20] Whatever its cause, his hesitation would haunt him a year later.

Not that Huygens had immediately forgotten the discovery upon announcing its companion to Sluse and van Schooten, nor that he had no warning of what might ensue. Almost immediately Huygens's priority was challenged, unwittingly it seems, by van Heuraet, with their mutual mentor, van Schooten, caught in the middle. Unfortunately, most of the letters that might have painted a precise picture of the dispute are lost, and consequently various interpretations of events are possible. However, an outline of correspondence is easily drawn from a lengthy reply by van Heuraet that summarizes the argument before declaring an end to it.

After receiving Huygens's description of his result regarding the reduction of the surface area of the paraboloid to a circle, van Schooten apparently wrote back saying that van Heuraet had made a similar discovery. The evidence shows that subsequently van Heuraet provided Huygens with an algebraic expression of the relationship between the surface of the paraboloid and its circular base.[21]

More general than the information sent by Huygens to van Schooten, van Heuraet's formula clearly indicates that he was not merely mimicking Huygens's idea.

Huygens, however, was quite jealous of his invention and obviously unwilling to grant van Heuraet any recognition. He asked, if not demanded, whether van Heuraet had managed to unearth his other discovery as well. Clearly Huygens did not disclose the result itself, for van Heuraet retorts in his summary that he could not judge whether their work coincided without first knowing what Huygens's "other discovery" was. To overcome this impasse, van Schooten, in the only other letter that survives, writes to Huygens:

However, concerning your other discovery about which you indicated nothing to me except some obscure things, which were almost completely forgotten by me when I visited Heuraet, so that I could not remember what it [your discovery] properly comprised, much less whether his method corresponded with that of your discovery. Wherefore, he has asked that you please construct some problem by which occasion he may indicate it [his method], by no means reluctantly, whereby he can satisfy your desire.[22]

The very next day, Huygens reiterated his demand, apparently revealing nothing more about his discovery than he had originally, while at the same time alluding to still other problems couched in the same vague terms. Uncertain how, or even if, to reply, van Heuraet (again, according to his own summary) was pressed two weeks later with yet another letter, in which Huygens required him to abjure claiming credit for any result that followed from van Heuraet's own technique if that result was found to correspond to one of Huygens's discoveries.

Although a flippant undercurrent seems to suggest that he was smarting from the bitter turn of events, van Heuraet's overall reply conveys his justifiable indignation in a remarkably mild tone:

If you had only known my character, it would not have been necessary to exert so much effort against me, who by no means shall seek to rob you of the pleasure and honor of the aforesaid invention, even if the same might have been found by me long ago.[23]

Van Heuraet urges Huygens to keep his discoveries to himself until he is ready under "public pressure" to publish them formally. Rather self-consciously, he closes "with a laugh" by appending a list of nonsense anagrams, thus mocking Huygens's preferred method of ensuring priority without publication. The matter was over, and having vehemently bullied van Heuraet into abandoning whatever priority he might have rightfully claimed, Huygens did nothing more, formally or informally, about circulating either of his discoveries. In particular, not even van Schooten learned, at that time, the exact nature of the "other discovery," which related the parabola's length to the hyperbola's area.

This state of affairs is surprising because in the extant letter van Heuraet frequently refers to his own "method" and quite clearly valued it above any solution it could facilitate with regard to Huygens's "particular problems." Yet Huygens appears not to have reacted to this evidence regarding a new technique. Of course, he could scarcely inquire into van Heuraet's method when he had just insinuated that the fellow was a plagiarist. Most priority squabbles in which he engaged ended with Huygens, and not his opponent, making the grand gentlemanly gesture of rising above the dispute and declaring it inconsequential. His embarrassing behavior, which van Heuraet's letter obviously spotlighted, probably reinforced his usual reluctance to publish (familiar even to van Heuraet), with the consequence that he declined even to establish his priority regarding the "other discovery" by telling Sluse or some other correspondent unaware of the tiff with van Heuraet. Another likely factor contributing to his lack of action at this time was his perpetual tendency to undervalue a general technique developed by another in favor of his own elegant, usually narrowly applicable solutions. For whatever reason, Huygens did nothing more with his results for an entire year.

In January 1659, however, he finally communicated his discovery regarding the relationship between the parabola and hyperbola to select correspondents, including Sluse and van Schooten. His receipt at the beginning of the month of Pascal's *Histoire de la roulette* with its wealth of discoveries and his subsequent generalization of Wren's rectification of the cycloid had, at last, spurred him into action. Huygens's genuine admiration of Wren's achievement was quickly followed by an apparent need to display his own mathematical talents, particularly in light of his failure to solve all of the challenge problems posed by "Dettonville," the more intractable ones of which he had even at one point declared to be insoluble.[24] Whatever had prevented him from circulating his own result before vanished, and he busied himself with reporting (after praising Wren's coup) both his generalization of Wren's result and his discovery connecting the rectification of the parabola with the quadrature of the hyperbola.

His announcement to friends in Paris brought the surprising reply that more than a month before Adrien Auzout had discovered the same relationship between the parabola and hyperbola.[25] His announcement to van Schooten, coupled with a complaint about Auzout's "coincidence," brought an even more disconcerting response, although Huygens should not have been caught unaware, given his previous correspondence with van Heuraet. Van Schooten revealed that, coincidentally, van Heuraet in his work the year before had also discovered the relationship and, moreover, had developed a general theory regarding the correspondence between rectifications and quadratures which subsumed that result as a mere corollary (!) and which van Schooten intended to publish in his second edition of the *Geometria*.[26]

Huygens's original announcement to Sluse, actually the first in the series, had engendered no response at all. But a second letter, intended to enlist Sluse as witness to his priority over Auzout, by citing his earlier correspondence, brought a disastrous reply. Sluse could not remember having received the first result regarding the paraboloid, let alone its vague reference to the "other discovery." Pricked by another reminder from Huygens, who repeated the exact contents of the paraboloid theorem, Sluse's memory improved.[27]

If the extant summary is a true reflection of its final content, Huygens's first announcement to Wallis, made in January along with the others, had contained only a vague claim to be able to evaluate the surfaces of conoids and spheroids.[28] With his priority threatened by Auzout and van Heuraet, however, Huygens became more explicit. The summary of the letter to Wallis in which he tells the Englishman of van Heuraet's work contains a remarkable claim:

When he [van Heuraet] learned that I had measured the surface of the parabolic conoid and had determined the length of the parabola equal to a given quadrature of the hyperbola (concerning both of which I wrote you previously), he found not only both of them by his own technique but, in addition, he rectified completely all other curves of those genera that we allow in geometry.[29]

Huygens repeats this account to Sluse a month later.[30]

No evidence exists to substantiate this claim. Indeed, everything in van Heuraet's letter to Huygens suggests that van Heuraet had come to his results independently but had ceded Huygens credit for the discovery regarding the measure of the paraboloidal surface in order to terminate a quarrel. Almost certainly, van Heuraet never learned that Huygens had related the length of the parabola to the area under the hyperbola, the infamous "other discovery" about which Huygens had *not* written either van Schooten in 1658 or Wallis in 1659 (as he claims to Wallis). Thus, just as certainly, van Heuraet could not have molded his new theory on that discovery.

Realizing his excess or mellowed by time, Huygens published a modified version of his claim in the history of rectification that appears in Part 3 of the *Horologium Oscillatorium*. After declaring van Heuraet's rectification of the semicubical parabola superior to Neil's, Huygens goes on to place himself first before all:

Indeed, since we are on this topic, please allow us to explain what we contributed to the promotion of so exceptional a discovery: because we provided to van Heuraet the occasion for having found it and we discovered the dimension of the parabolic curve from the given quadrature of a hyperbola (which is part of what Heuraet discovered) before him, indeed, first of all. For at the end of 1657 we discovered two things simultaneously, namely, the dimension of the parabolic curve of which I have spoken and the reduction of the surface of a parabolic conoid to a circle. And when we indicated this by letter to van Schooten, as well as to other friends, stating that we had found two unusual things regarding the parabola and that one of them was the transformation of the conoidal surface into a circle, he shared that letter with van Heuaret, with whom he then associated. Indeed, it was not difficult for this man of very keen ability to deduce that the surface of that conoid is associated with the measure of the parabolic curve itself. Having found both, the investigator then moved on to those other paraboloidal curves for which straight lines equal to their length can be found absolutely.[31]

It is entirely possible that Huygens truly believed that his account was accurate history; indeed, it is entirely possible that this final version is true. However, without van Schooten's first letter, in which he would have broached the subject of van Heuraet's work, it is impossible to ascertain whether van Heuraet did draw on the result regarding the paraboloid that was communicated.[32] Van Heuraet's wording and the intensity of the priority squabble seem to point to independent discovery. On the other hand, Huygens's editors would argue, "Huygens assures us, and there is no reason to doubt

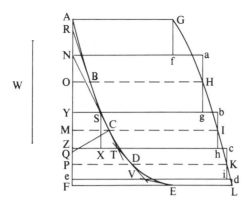

Figure 7.1. Van Heuraet's theorem.

him," that van Heuraet's work was inspired by van Schooten's report of Huygens's original letter.[33] Of course, van Heuraet and van Schooten could scarcely contradict his story because, by the time the *Horologium Oscillatorium* was published in 1673, both had been dead for more than a decade.[34] Unfortunately, there is no way to track the development of van Heuraet's contributions barring a discovery of his unpublished papers.

All that remains of van Heuraet's theory is the outline published in the second edition of van Schooten's *Geometria,* where that account itself is reproduced from a letter sent by van Heuraet to van Schooten in January 1659.[35] The method was a notable achievement, certainly the most important contribution to the mathematics of rectification to date, and it immortalized van Heuraet in the history of mathematics, just as van Schooten had predicted to Huygens.[36] In a few succinct pages, van Heuraet outlines a uniform method for transforming the rectification of a curve into the quadrature of a companion curve.

For any point C on the curve $ABCDE$ (Fig. 7.1) van Heuraet constructs a point I on a second curve $GHIKL$ according to the proportion $MI/W = CQ/CM$, where W is a constant and CQ is the normal at C. By the similarity of right triangles, $CQ/CM = ST/SX$ $(= ds/dx)$, where ST is the tangent at C. Whence $MI/W = ST/SX$, and $MI \cdot SX = W \cdot ST$; that is, the area of the rectangle of length MI and very small width SX equals the constant times the infinitesimal arc length ST. Increasing the number of rectangles ad infinitum and summing all their areas yields the conclusion that the area under the curve *GHIKL* is equal to a constant times the length of *ABCDE*. Specifically, van Heuraet shows that the semicubical parabola $ay^2 = x^3$ has as its companion curve the parabola $z^2 = \frac{1}{4}ax + \frac{1}{9}a^2$, whose quadrature can be computed, thanks to Archimedes.[37] In other words, as Neil had already discovered, in the particular case of the semicubical parabola, because the problem can be transformed into one with a known solution, the arc length can actually be determined. Van Heuraet goes on to demonstrate that the rectification of the parabola $ay = x^2$ reduces to the quadrature of the hyperbola $z^2 = 4x^2 + a^2$. He had indeed found Huygens's "other discovery."[38]

Even if Huygens's claim of influence has validity, his own contribution was modest in comparison. He had discovered the relationship between rectification and quadrature in the instance of the parabola and hyperbola, but he had done nothing more with it. In fact, it was quite clearly a sidelight to what he considered the important result, the reduction of the paraboloidal surface to a circle. Huygens's research in the period immediately following his discovery of the two results was occupied by his generalization of the latter proposition to other conoids. He did not attempt to apply his knowledge of the interrelationship between the rectification of the parabola and the quadrature of the hyperbola until much later.

In the fall of 1661, while pursuing an idea presented by Grégoire de Saint-Vincent in his book on the quadrature of the circle, Huygens represented logarithms by areas under the hyperbola and thereby gave them geometric justification. Nearly a year later, he reversed the procedure and numerically evaluated the arc length of a parabola by calculating the logarithm of the corresponding area under the hyperbola.[39] Naturally, these related discoveries are reported in the *Horologium Oscillatorium*.[40]

EVOLUTES AND RECTIFICATION

For the time being, in late 1659, in the flush of creative activity spawned by his work on gravity and centrifugal force, Huygens was developing his theory of evolutes and waiting for the publication of the two most important treatises yet produced concerning the problem of rectification. In England, Wallis was preparing his account of Neil's and Wren's work, and nearer to home, van Schooten was sending van Heuraet's letter to press.[41]

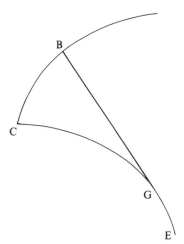

Figure 7.2. Rectifying an evolute.

Given the apparent pattern by which Huygens chose research topics, it is not at all surprising that he should have concentrated on the question of rectification rather than curvature when he turned to the mathematical application of his theory.[42] It was a prestigous problem "not solved by our forefathers." The goal was in sight, with the work of Neil, Wren, and van Heuraet pointing the way. Moreover, although his own work on the arc length of the cycloidal segment and the reduction of the paraboloid had been important contributions, they had not been so vital as to keep his name at the forefront of the field. He needed to press forward.

Fortunately, the application of evolutes to the clock placed the solution of the problem of rectification squarely in front of him and, in fact, made it nearly impossible for him to follow any other path. As the cord of the pendulum unwinds off a curved plate, it literally straightens, or rectifies, that curve, and the freed portion of the cord by the very nature of its generation equals the arc of the plate from which it has been unrolled. It is a commonplace method for physically measuring irregular or curved surfaces, a method embodied today in the tape measure. Huygens's method of rectification is nothing but its mathematical counterpart.

In the clock's mathematical analog, line *BG* (Fig. 7.2) equals arc *CG* because it corresponds to the unwound portion of the curve *CGE*. Every time Huygens derived an evolute by applying his fundamental proportion, he calculated the length *BG*, because his basic method required him to find, for each point *B* on the involute, the position of the corresponding point G on the evolute. He first determined G's position relative to B by computing its distance BG along the normal to the involute at B. Only then did he derive the equation of the evolute with respect to a fixed origin. Thus, every derivation of an evolute was simultaneously a rectification of that evolute.

So natural was this process that Huygens apparently felt no need to formalize his rectification technique into a separate set of theorems. However, although he created no independent theoretical framework for the use of evolutes in rectification, he did explicitly refer to that application, in contrast to his complete silence on curvature, which to a modern observer seems an equally natural application. His outline for the expanded version of the *Horologium* of 1658, made immediately after his development of evolutes and their application to the cycloidal clock, refers to his technique for rectifying curves and to a plan to contrast his method with van Heuraet's.[43] And a few pages later his workbook contains his only formal statement regarding rectification, a "treatise" only one paragraph long stating that evolutes can always be derived for geometric curves and that, consequently, those evolutes "can be reduced to straight lines."[44] Derivations of the arc lengths of evolutes to higher-order conics follow, and as would be expected, this collection of results duly appears in the published *Horologium Oscillatorium*.[45]

Even in the *Horologium Oscillatorium* Huygens does not separate rectification from the determination of the evolute itself. The final theorem of Part 3 essentially repeats the statement of the workbook:

Proposition 11

Given a curve, to find another by whose evolution the first is described; and to show that for any geometric curve whatever, another curve, likewise geometric, exists, to which a straight line can be equated.[46]

The presentation of his general method for deriving an evolute from the fundamental proportion follows and includes a justification for introducing his equivalent to the second derivative, thus making the evolute "likewise geometric."

Of course, the first curve that Huygens had measured had been the cycloid. However, he had not applied his new technique in the original derivation; rather, he had drawn upon his intimate knowledge of Archimedean methods to generalize Wren's result. In fact, it was because he knew the arc length of a cycloidal segment by an independent means that he was able to derive his first evolute, for the cycloidal involute. Nevertheless, the proof was totally reversible, and as soon as he had the general theory of evolutes, he could rederive the arc length of the cycloid using the new technique. It is in this mode that the rectification of the cycloid is presented in the *Horologium Oscillatorium*.[47]

Huygens emphasized the universal applicability of his method by providing rectifications, redone according to his own technique, for all the important geometric curves of the seventeenth century. Proposition 11 of the *Horologium Oscillatorium* concludes with a list of involutes and their corresponding evolutes that includes the cycloid, the semicubical parabola, the standard conics, as well as many higher-order conics.

This was a very great and enviable achievement. Wren and Neil had found the arc length of only one curve each, and neither had tried generalizing his procedure. Van Heuraet had rectified any general curve but only by replacing the problem of length by a problem of area, which brought with it no guarantee of solvability. Van Heuraet's achievement was very similar to Huygens's generalization of Archimedes' quadrature of the parabola in *Theoremata de quadratura,* in which Huygens transformed quadratures of the conic sections into problems involving centers of gravity. These transformation methods were valuable because they gave mathematicians new avenues by which a solution might be broached and also because they provided researchers with a sense of the underlying structure by which mathematical techniques are related. It is in the latter vein that van Heuraet's theory is remembered today, as a precursor of the fundamental theorem of the calculus.

Yet how much more valuable is a universal method that can yield an answer directly. It was on this very point that Huygens had praised Wren's measurement of the cycloid as "excellent, because it is the first and perhaps only curve that can be rectified."[48] And it was because his own method produced a straight line that his technique outweighed van Heuraet's transformation.

His pride in his discovery was freely displayed in the letter of December 6, 1659, in which he informed van Schooten of his improved clock. Without actually disclosing his method, Huygens told van Schooten that he was able to determine the precise shape of the curved plates "by geometric reasoning" and concluded, "I venture to say that the discovery will not displease the very discriminating van Heuraet; for to me, certainly, it appears to be the most felici-tous of all the discoveries that I have ever hit upon."[49] Presumably, the reference to van Heuraet was a challenge to that mathematician to duplicate Huygens's result. Obviously, Huygens had not forgot-ten the earlier tiff, for his remark intimates that this time – unlike the last, to Huygens's way of thinking – van Heuraet would be un-able to replicate his discoveries.

The achievement was Huygens's alone. Not only was the arc length directly and universally attainable, it was dependent upon a technique that stemmed from his greatest mechanical invention. The same interdependence between the mathematics of evolutes and the physics of pendulums that fueled his creative process also guaranteed the uniqueness of the finished product. The reputation of both Huygens the mathematician and Huygens the mechanic was enhanced, a fact of no little consequence. With his new method at hand, Huygens could sidestep the whole controversy of priority and dependency, not just with regard to rectification, but also with respect to the pendulum clock.

PRIORITY AND THE PENDULUM

Another challenge to his reputation had been advanced during the same period in which Huygens was claiming to have influenced van Heuraet's study. The publication in 1658 of his original *Horolog-ium,* in which he had introduced the simple-pendulum clock to the general public, had precipitated a series of priority claims concern-ing both the idea of a pendulum clock and the mechanical realiza-tion of that idea. In the spring of 1659, having already defeated at-tempts of his clockmakers to share the patent, Huygens was faced with the accusation that, first among others, Galileo had invented the pendulum clock. The formal complaint came from none other than Prince Leopold de' Medici, out to protect the memory of his family's resident genius:

Concerning the clock regulated by a pendulum, certainly the invention is beautiful, but one must not steal any of the glory due to our forever admirable Galileo, who already in 1636, if I am not in error, proposed this same useful invention to the States General of Holland, and I have rediscovered a model already built by the same Signor Galileo, although in part different in the matter of the makeup of the gears.[50]

The complaint had been sent to Ismael Boulliau, who wasted no time in defending Huygens's honor. Forwarding Huygens a copy of Leopold's accusation, Boulliau summarizes his own reply:

I have responded on this matter to His Most Serene Highness that I know that you would consider it an honor (and that you believe to merit the glory) if you had fallen upon the same thoughts as Galileo had made; and that you are so much a man of honor and so sincere that you would never rob from the reputation of another in order to attribute it to yourself; that you have an extraordinarily fertile mind for very beautiful inventions, and thus in order to satisfy yourself and in order for you to acquire renown you do not need the inventions of another.[51]

After expressing his thanks for the kind words, Huygens tells Boulliau in return that no one seems to know of any horological invention presented by Galileo to the States General of Holland except the Italian's proposal (made in 1636) to use the moons of Jupiter to determine longitude. He does not deny that Galileo might have conceived of a clock regulated by a pendulum, however. Indeed, everybody seems to be claiming priority of invention over him, including Roberval, Hevelius, and Wallis (on behalf of Englishmen everywhere): "So that the thought seems to have been very common, but one cannot deny that my model was the first to succeed."[52] Continuing, he bemoans the fact that, since he is not known well enough at the Medicean court to be able to ask a favor, he cannot request a diagram of Galileo's apparatus and thereby see how that clock differs from his own. A difference only in the gears is nothing significant, he concedes, but a difference in the mounting of the pendulum (which he obviously suspects is the case) can undermine the success of the device. Defending himself against the charge of plagiarism, he asserts:

Certainly, I would believe me undeserving of life, but since, however, the negative is difficult to prove, I do not see what other apology I could make to His Most Serene Highness except to protest to him with all sincerity that neither I nor anyone in this country, whom I have been able to consult, heard this invention discussed before I created it.[53]

Huygens promptly arranged to dedicate to Prince Leopold his *Systema Saturnium,* which was then on its way to press.[54] Undoubtedly, he was hoping his revelation that Saturn was surrounded by a ring (something Galileo saw but misinterpreted) would prove his ingenuity and convince the influential Medici to think more favorably of him.

In the meantime, Leopold had responded to Boulliau's defense of Huygens by explaining to Boulliau that he never had wished to imply that Huygens could not have discovered the pendulum clock independent of Galileo. "Knowing very well the Eminence of his Genius and thought," the prince concludes, "that great Virtuoso" is definitely capable of creating such a device and even greater things.[55] Obviously, Huygens was much more well known, and favorably so, at the Medicean court than he had reckoned.

Although he took delight in Leopold's unexpected praise, Huygens was still faced with the question of priority.[56] Acquitted of stealing the idea, he was not himself to be robbed of the acclaim for inventing the first accurate pendulum clock, no matter that his rival for the honor was his hero Galileo. Thus, the dedication made and the compliment received, Huygens waited for a reply from Florence. What was Prince Leopold's reaction to the *Systema Saturnium?* Would Boulliau, who had quickly grasped his hint, be able to arrange for Huygens to inspect a diagram of Galileo's clock? In the fall of 1659, Huygens waited for the answers that would confirm his reputation in horology, just as he was awaiting the publications by Wallis and van Schooten that would determine his standing among mathematicians on the problem of rectification.

Yet the results that he achieved while he waited perforce invalidated whatever judgment might have been rendered. His new clock was, theoretically, absolutely precise, something Galileo's pendulum could never be, bound as it was to its nonisochronous path. His theory of evolutes could, in principle, measure the length of a curve directly, unlike van Heuraet's transformation, tied as it was to another problem that might not itself be solvable. Definitely, the revised verdict could only be that Huygens was still the master of masters.

The verdict of Leopold and his court was less satisfying. Although Leopold told Boulliau that the *Systema Saturnium* confirmed the fame of its talented author, he told Huygens nothing, much to Huygens's continuing annoyance.[57] A misunderstanding on Huygens's part of Florentine court etiquette delayed a direct response by the book's dedicatee for more than a year, by which time those in Italy who opposed Huygens's ring hypothesis had taken the opportunity to voice their objections to the prince.[58] Instead of overwhelming Leopold with his brilliance, Huygens became embroiled in still another dispute, with Leopold as judge and the Accademia del Cimento as jury.[59]

The pendulum controversy slowly sank from view, raising a few intermittent waves before disappearing entirely. Leopold did send Boulliau drawings of both Galileo's design and the model extant in the Medicean palace, as well as a detailed history of Galileo's work on the pendulum written by his pupil, Vincenzio Viviani.[60] A few months later, after making copies for himself, Boulliau forwarded the drawings to Huygens but kept back Viviani's history, saying nothing about this additional evidence. Leopold had instructed him to show the history to a friend of Galileo still alive in France, and since Boulliau did nothing further with the work beyond this task, although he had earlier assured Leopold that he would send a copy to Huygens, it is not too much to presume that the friend's recollections prompted Boulliau to devalue Viviani's scenario. There is no record that Leopold learned of either the friend's reaction or Boulliau's change of plans, nor was Leopold disposed to ask how Huygens had reacted to the history. Viviani's tale was ignored, undergoing only a brief revival of interest when the *Horologium Oscillatorium* appeared.[61]

The drawings that Boulliau had forwarded to The Hague afforded Huygens the opportunity to restate his claim that only his clocks were accurate and reliable machines. He tells Boulliau that Galileo's design was inferior because the standard crown wheel had been replaced by a "device much more difficult." Undoubtedly to his great satisfaction, he could report that with its solid arm mounted directly onto the shaft Galileo's pendulum would be susceptible to distortions and even to stopping, as he himself knew by experiment. In this manner, Huygens implies that he too had thought of Galileo's design but had rejected it as unworkable. Although both he and Galileo had thought of applying the pendulum to a clock, he continues, the difference in their final designs at least verifies his claim of independent discovery and absolves him from the charge of plagiarism. Having proclaimed the independence and even superiority of his original design, Huygens concludes with the announcement to Boulliau regarding his plans for a new edition of the *Horologium* with a more advanced clock (the cycloidal model) based on the science of accelerated motion first treated by Galileo.[62] Galileo was still to be revered, it seems, just not at Huygens's expense.

Regarding this alternative history of the pendulum, neither concurrence nor denial was forthcoming from Florence. A year later, Huygens, by then in direct communication with Prince Leopold, appended a description of his cycloidal clock onto his latest response regarding the ring controversy.[63] His newest advance in horology was tendered, presumably, as proof of the keenness of the mind that had conceived of a ring around Saturn. In this courtly passepied, Huygens's new pendulum would now be used to substantiate the book that had previously been tendered as collateral for the original pendulum designs.

Over the course of all these controversies, Huygens's reputation survived, nay prospered. His *Systema Saturnium* was judged by the Accademia del Cimento as the hypothesis most likely true, and his cycloidal clock became the rage of scientific fashion in Paris and London. Only his mathematics of rectification remained relatively neglected, not through any lack of appreciation by the scientific community but because Huygens did not publish his results for another thirteen years. In the meantime, the mathematics remained cloaked within the physical entity that had provided fertile ground for its creation. However, any mathematician of talent who had knowledge of the cycloidal clock could attempt his own proof of the inherent mathematics. The clock itself guaranteed Huygens's priority in all such matters, allowing him time to prepare his own meticulous analyses of its mathematical structure for publication.

DIMENSION REDUCTION AND PI

The collection of propositions eventually published in Part 3 of the *Horologium Oscillatorium* encompasses more than the mathematics necessary to explain the curved plates of the cycloidal clock. Amid the theorems concerning evolutes is a set of problems that for a modern reader comprises undoubtedly the most inaccessible and antique section of the book.[64] Yet quite clearly Huygens and his contemporaries valued these propositions, which deal with the reduction of three-dimensional surfaces to planar figures, just as highly as those related to evolutes and rectification. Huygens included them in the *Horologium Oscillatorium* because, as was the case with so many other propositions in that book, their creation as well as their content was intimately tied to the propositions that did apply to the clock.

All the purely mathematical propositions of Part 3 have a common heritage traceable to the pair of results concerning the parabola that Huygens discovered in 1657. Whereas the propositions regarding the application of evolutes to the problem of rectification are indirectly related to Huygens's proof that the arc length of the parabola can be determined by the area under a hyperbola, the propositions on surface reduction are directly linked to Huygens's transformation of the surface area of a paraboloid to the area of a circle.

The background of the two results themselves is difficult to delineate, because the manuscript evidence is sketchier and thus more malleable than that from Huygens's fertile period two years later. Certainly, the aforementioned claim that Hobbes's polemic provoked Huygens's research is one possible explanation of why Huygens tried to rectify the parabola. But a more immediate, if less direct, stimulus for his discoveries can be traced to the correspondence between Huygens and Sluse during the second half of 1657.

Among the many mathematical topics Sluse raised for discussion was the curve now known by his name, the Pearl of Sluse. Sluse challenged Huygens to find the area inside the curve, its center of gravity, and its tangent at any point. Huygens obliged within the month and also passed the problem on to van Schooten, who in turn shared it with his students, van Heuraet and Johan Hudde.

The three-way correspondence thus initiated became a discussion of general methods for analyzing similar curves, which in some sense can be pictured as distorted or generalized parabolas.[65] These methods contributed greatly to the foundations on which the calculus was later to be constructed. Sluse, Huygens, and Hudde each developed techniques for determining tangents and finding maximums and minimums. Hudde's rule, appended to van Schooten's 1659 edition of the *Geometria,* became the most famous protodifferentiation method of the seventeenth century. The other appendix to the *Geometria* that rivaled Hudde's in its importance to the rise of the calculus was none other than van Heuraet's transformation theorem.

Although Sluse never asked his correspondents to determine the length of a parabola, both Huygens and van Heuraet discovered its relationship to the area under the hyperbola at this time. Indeed, Huygens's announcement to Sluse that he had found two results concerning the parabola accompanied his latest report of the activities of the Leiden mathematicians.[66] Surely, the simultaneity of discovery in this case is attributable to the common stimulus of Sluse's desire for a mathematical dialogue regarding generalized parabolas, even if the exact nature of his influence cannot be traced to a specific problem or comment.

Poor Sluse, who was at the mercy of his correspondents, especially Huygens, for news of the mathematical world, was often neglected. As late as August 1660 he could be found complaining that, among other treatises, he still had not seen van Heuraet's work, even though he obviously had a vested interest in the treatise, which at that point had been in circulation for nearly a year.[67] As already shown, Huygens himself was not particularly forthcoming with Sluse unless it suited his own needs. Consequently, a year passed before Sluse learned from one of its discoverers the relationship between parabola and hyperbola that he had probably unwittingly inspired and that had caused so much ill will between the Dutchmen, the infamous "other discovery."

Surviving on the meager crumbs allotted him, Sluse, when told by Huygens of the latter's reduction of the paraboloid to the area of a circle, generously gave praise:

I rate all these [other] curves and even the entire linear locus itself [a technique Sluse was developing] as insignificant, in comparison with your discovery, by which you determined the ratio of the surface of the parabolic conoid to the circle at its base. I willingly prefer this very beautiful *reduction* for the quadrature of the circle to all those many that I formerly deduced from the linear locus and that, given the occasion, I will communicate to you if you so desire.[68]

Huygens repeats this passage verbatim in his history of rectification for the *Horologium Oscillatorium,* presumably because Sluse's comment provides proof of his priority regarding the result as well as an evaluation of its worth.[69]

Sluse's enthusiasm sprang from his hope that Huygens's transformation could solve that intractable problem of geometrically measuring the area of a circle. Instead of reducing the area to a square of determinable side length, Huygens's transformation would find the quadrature of the circle indirectly, by determining the surface of a paraboloid through some independent means as yet unknown.

This connection between Huygens's results of 1657 and the problem of squaring the circle might be interpreted as evidence of Hobbes's influence on his work after all, albeit by a different path than that usually claimed. However, it seems much more reasonable to view Hobbes's erroneous quadrature of the circle as evidence of the problem's persistent hold on the imagination of seventeenth-century mathematicians, Grégoire de Saint-Vincent's attempted solution and Sluse's endorsement of Huygens's alternative being further manifestations of this widespread interest.

Although clearly rooted in Greek geometry, the problem of squaring the circle was not merely a reflection of the contemporary penchant for reviving ancient learning. Even those savants of the seventeenth century who had no ambition to reconstruct ancient mathematics needed to determine the fundamental mathematical constant, pi. Defined as the ratio of the circumference of a circle to the diameter, pi can be expressed equivalently as the ratio of the area of a circle to one-fourth the square of its diameter. Thus, modern pejoratives aside, to square the circle is to evaluate pi.

As mathematicians moved further and further away from pure geometry and toward analysis in the seventeenth century, the classical definition of pi became an awkward remnant of ancient requirements that relationships be expressed in proportions. Recall Huygens's use of p/q in his derivation of the isochronism of the cycloid when he reached the stage where he wanted to express his equation algebraically. Remember as well that π is only a symbol, of no more inherent value than p/q (and not standardly used until the eighteenth century).

Of course, attempts to evaluate pi were scarcely new. Indeed, any society with enough mathematics to contemplate the question seems to have assigned pi some value, however imprecise. In particular, Archimedes' geometric technique of approximating the circle with inscribed and circumscribed polygons in order to determine lower and upper bounds on pi's value had generated many similar treatises, each successively narrowing the gap between bounds and thus improving the approximation, including Huygens's own refinement in *De Circuli Magnitudine Inventa*. In the middle of the seventeenth century, algebraic techniques for evaluating pi became common. Wallis's infinite product,

$$\frac{4}{\pi} = \frac{3 \cdot 3 \cdot 5 \cdot 5 \cdot 7 \cdot 7 \cdot 9 \cdot \ldots}{2 \cdot 4 \cdot 4 \cdot 6 \cdot 6 \cdot 8 \cdot 8 \cdot \ldots},$$

and Brouncker's continued fraction,

$$\frac{4}{\pi} = 1 + \cfrac{1}{2 + \cfrac{9}{2 + \cfrac{25}{2 + \cfrac{49}{2 + \cdots}}}},$$

were eventually joined by Leibniz's infinite sum,

$$\frac{\pi}{4} = 1 - \frac{1}{3} + \frac{1}{5} - \frac{1}{7} + \cdots.$$

All these techniques for delimiting pi, be they geometrically or algebraically garbed, involved repeated processes that were considered to proceed ad infinitum but were truncated at some arbitrary point for evaluation purposes.[70] Of course, today we know that any method for determining pi will inevitably involve an appeal to the infinite, because pi is transcendental. However, in the seventeenth century the question was still open, despite an attempt by James Gregory in 1668 to resolve it.[71] Perhaps there was some approach to the problem of squaring the circle whereby one could avoid those infinite processes that, once truncated, could only yield an approximation. This was the promise contained in Huygens's transformation, which substituted the paraboloid, whose progenitor had always been the most amenable of curves, for the unruly circle.

Although the debate with Grégoire de Saint-Vincent and his followers might suggest otherwise, Huygens did hope that the quadrature of the circle was attainable. His concern, as he tells one of Grégoire's defenders, is that the proof be convincing: "Truly, I have neither discerned nor prescribed by what means a circle may be squared; but I urge this, that he who contends to have found that means should demonstrate it to be in truth useful and effective."[72]

The contemporary interest in squaring the circle may well explain why, once Sluse's inquiries had led him to the two results of 1657, Huygens emphasized his reduction of the paraboloid to a circle rather than the transformation of the arc length of the parabola to the quadrature of the hyperbola, which is the result most valued today. Indeed, immediately upon transforming the parabola's length to its hyperbolic counterpart, Huygens realized that the paraboloidal result followed, and thereafter he worked on the expansion of the second result.[73] Thus, a short while later, Huygens announces to Sluse (while chiding him about the gaps in his latest method for squaring the circle):

I indicated to you before that the surface of a parabolic conoid can be reduced to a circle. Now, truly, I know that the quadrature of the Circle is determined, if we can find a circle equal to the surface of a prolate spheroid. Likewise, the quadrature of the Hyperbola [is determined], if a circle may be made equal to the surface of an oblate or compressed spheroid; or to the surface of a Hyperbolic conoid. And contrarily.[74]

A month later, when Sluse erroneously determined the area of the circle by means of the cissoid, Huygens responded with a derivation that reduced the unbounded area between the cissoid and its asymptotic axis to three times the area of a semicircle.[75]

Huygens also communicated both sets of results to Wallis, who became the principal agent for their dissemination. Wallis published Huygens's proof of the cissoid result, as well as his own version of the theorem, in his *Mechanica* (1669–71) and gave his own derivations of the conoid transformations in *Tractatus Duo* (1659).[76] Thereby relieved of any pressure from his friends to publish his work independently, Huygens abandoned the tedious process of preparing finished proofs of his discoveries.[77]

A decade and a half after their genesis, all these mathematical insights attendant on the development of evolutes finally appeared in the *Horologium Oscillatorium,* wedged into the middle of Huygens's explanation of evolutes themselves. His rectification of the semicubical parabola by means of evolutes (Proposition 9) is followed by his history of earlier rectifications, in which the contributions of Neil and van Heuraet are acknowledged. Raising van Heuraet's name gives Huygens the opportunity to make his claim that his own work on the paraboloid influenced van Heuraet's discovery, and the accompanying quotation of Sluse's praise corroborates his dating, if not his priority. The excerpt from Sluse, in turn, introduces the problem of the quadrature of the circle, and an outline of Huygens's transformation theorems readily follows: The surface areas of the paraboloid, oblate and prolate spheroids, and hyperboloid can all be reduced to the areas of appropriately related circles (as well as to each other). Finally, Huygens returns to rectification, showing that the parabola's length is related to the hyperbola's area, which is itself reducible to logarithms. Only then does the topic return to evolutes per se, as Huygens proceeds to derive the evolutes of ellipses and hyperbolas (Proposition 10).[78]

The juxtaposition of his presentation of the conoid discoveries and his claim to have inspired van Heuraet is a pathetic example of seventeenth-century scientific jealousy. Whereas Huygens's influence on van Heuraet is difficult to substantiate, van Heuraet's influence on Huygens's approach to the conoids can be documented, although it certainly never was acknowledged by Huygens. As a sign of good faith during their priority dispute, van Heuraet had revealed to Huygens (via van Schooten) his own algebraic expression for the ratio of the surface area of a paraboloid to the area of its circular base.[79] Obviously desiring to check van Heuraet's work against his own version, Huygens immediately, and successfully, tried to derive van Heuraet's formula. The technique that he used in this derivation was quite different from his original approach and, in particular, was much more manageable. As a consequence, Huygens was able to extend the concept of surface reduction to the other conoids, and he quickly reduced their surface areas first to circles and then, alternatively, to hyperbolas.[80] When Huygens alluded to another set of problems that, along with the pair concerning the parabola, van Heuraet was to renounce priority over, it was undoubtedly this very group of results. A month after van Heuraet had communicated his formula to van Schooten, Huygens was telling Sluse of his new discoveries with nary a word about their source of inspiration. Huygens's behavior toward van Heuraet reflected the typical procedure among researchers of his day: Claim priority on the slightest evidence and ignore obligation whenever possible.

Huygens was never able to evaluate any of his alternatives to the quadrature of the circle, but that fact should not be construed as a sign of failure. At a minimum, the transformations themselves added to the general knowledge of mathematical structure, just as his reductions of the quadratures of the planar conics to their centers of gravity had done. Moreover, if pi could be found, by whatever means, a whole class of quadratures would be determined immediately. That is, Huygens had geometrically solved the problem of finding the area of a surface of revolution. Because, in its algebraic manifestation, the answer involves pi, the geometric solution quite naturally involves the area of a circle.

Put in more modern terminology, Huygens's transformations reduce two-dimensional surfaces embedded in three-dimensional space to their planar equivalents. This reduction parallels completely his rectification process, whereby a one-dimensional curve embedded in two-dimensional space is given its linear equivalent.

In transforming his problems to simpler dimensions and to standard curves, Huygens was conforming to the mathematical aesthetics of his day. H. J. M. Bos, outlining some of the characteristics of late-seventeenth-century and early-eighteenth-century mathematics, cites examples from the works of the Bernoulli brothers, Leibniz, and Huygens, in which curves are constructed by means of the arc lengths of simpler curves, so that problems of area become problems of length. Although rectification may not be inherently better than quadrature to modern eyes (because both involve the integral), Bos argues that it appeared so to these mathematicians:

But, as Leibniz wrote to Johann Bernoulli, it is better to reduce quadratures to rectifications because "the dimension of the line is simpler than the dimension of the plane". And often we read that the reduction of quadratures to rectifications is useful because arc-lengths are more easily measured than areas: one can take a chord and stretch it along the curve, thus measuring its length.[81]

One of the criteria for evaluating geometric simplicity seems to have been the existence of a physical referent, such as the ability to measure with a cord. This "idea of practical applicability," as Bos calls it, is epitomized by his final example, in which Jakob Bernoulli argues for the reduction of all constructions to the catenary, rather than the hyperbola to which it is related, because the chain curve occurs naturally without need of a mathematician to create it.

Both aspects cited by Bos are certainly evident in Huygens's mathematics. Huygens leaves the impression that to have simplified, particularly through dimension reduction, was in some sense to have solved a problem. And the physical analog of his rectification technique, namely the swinging pendulum of his clock, was as natural as Bernoulli's hanging chain.

Bos states an obvious, but no less important conclusion:

In representing curves the seventeenth century mathematician applied neatness requirements which originated in geometrical, almost mechanical considerations, and which therefore were quite different from the neatness requirements which the mathematicians of later times made in the representation of curves.

And this different context, this different way of conceiving mathematical objects and operations, had a considerable influence on the direction of mathematical investigation in the later seventeenth century.[82]

GEOMETRY AND THE CALCULUS

Nowhere is the distinctive conceptual framework of seventeenth-century mathematics more evident than with regard to the fundamental theorem of the calculus or, rather, with regard to the absence of the fundamental theorem as it is conceived today. The problems inherited, the techniques used, the aesthetic requirements imposed – all combined to direct researchers, including Huygens, toward a mathematical structure askew from the modern one.

Mathematics was still dominated by geometry, which had been revitalized by the recovery of accurate classical texts and the application of the new analytic methods of Descartes and Fermat. For the most part, the new algebraic expressions represented in a modern language the problems bequeathed by antiquity: Find the tangent, area, volume, length. The tangent problem became the foundation of the modern fundamental theorem very late in the century, when the search for its inverse developed as a primary problem (namely find a curve whose tangent at any point is defined by a given relationship). However, until then, the major distinction of the tangent problem was the fact that all of the others entail the determination of a dimensional quantity. Of course, in a modern approach those other problems involve the integral and therefore have a kind of equivalence in the calculus, but because this commonality of operator was not apparent in seventeenth-century mathematics, the problems were considered intrinsically different. Rather than focus on the latent inverse relationship between them and the tangent problem, mathematicians sought the correspondence between allied problems of dimension.

Because geometry is basically a holistic vision in which the finished curve takes precedence over the generative process, this emphasis on dimensional measure should not be surprising. Certainly, curves were more often than not defined pointwise (for example, the circle is the locus of all points equidistant from a given point), but the idea of an instantaneous event was not fundamental to mathematicians in the seventeenth century; indeed, it was still scarcely admissible. Yet the acceptance of the instantaneous, however imprecisely defined and inaccurately applied, is at the root of the calculus and its fundamental theorem, for the calculus is a language describing change, not measure.

This is not to deny the prominence of Cavalierian indivisibles and other infinitesimal methods as heuristic devices. The techniques eventually codified in the calculus were being developed and applied, but for the most part the emphasis remained on the depiction of the entire figure and not on the fluctuating parameters of a curve. Huygens's application of evolutes to rectification rather than curvature is a good example of this tendency to focus on geometric measure over analytic description.

Consequently, although the seventeenth century could be said to have had a fundamental theorem, it was one markedly different from that which relates the operations of differentiation and integration. Instead, the inheritance bequeathed by the Greeks focused attention on the correlation between the measures of quadrature and rectification. Archimedes had even explored this correspondence when he showed that the area of a circle is directly proportional to its circumference, thereby linking the problems of area and length in that quintessential paradigm. The statement of the general principle relating measures and thus the fundamental theorem of seventeenth-century mathematics was van Heuraet's transformation theorem linking arc length to the area under a curve.

Of course, van Heuraet's theorem is usually depicted by general histories of mathematics as a precursor of *the* fundamental theorem, falling short of the ultimate goal only because he represented the underlying relationship between tangent (used in his construction) and area geometrically. Yet it is very doubtful that van Heuraet grasped the inverse nature of the mathematical processes at hand. Moreover, although he did generate his new curve pointwise from the given one by means of the tangent, the latter was only a temporary device necessary for the derivation of the resultant curve. As with Huygens and evolutes, the scaffolding came down once the primary goal had been achieved, and van Heuraet's conclusion is about arc lengths and areas.

Usually paired with van Heuraet in histories of the calculus and similarly condemned for his attachment to geometry, Isaac Barrow came a little closer perhaps to understanding the operational nature of transformations. If nothing else, he at least unambiguously stated the relationship between tangent and area. Yet even his work stands squarely in the tradition of transformation of measures, for his theorem relating tangent and area is only one of many pairings that survey the interdependency of length, volume, center of gravity, area, and tangent.[83] Obviously, Huygens's two discoveries regarding the parabola and his subsequent generalization of the one into a formula for transforming surface areas of conoids into circles also lie firmly within this tradition.

As the depiction of van Heuraet's and Barrow's work exemplifies, geometry of the late seventeenth century has often been castigated as having somehow obstructed progress toward the calculus. In this view, the mathematical phase of the Scientific Revolution consisted of the overthrow of geometry by pure analysis: a new mathematics for a new physics. An extreme variant of this attitude argues that this revolt is still another example of the new Newtonian science defeating an inadequate Cartesian model, particularly that model provided by Descartes's devout follower, van Schooten, in his *Geometria*. As a consequence of his teaching, his pupils – none other than Huygens, van Heuraet, and Hudde – were "shackled" to a narrow, unproductive approach. The ultimate charge is that Huygens could, nay should, have derived the fundamental theorem while working on a problem similar to one that inspired Newton to achieve the goal but that Cartesian patterns of mathematical thought inhibited his solution.[84]

In applying his formula for evolutes to the conics, Huygens discovered that the evolute of an equilateral hyperbola is a sixth-degree curve. In the margin next to this derivation, he reworked his solution into the general form of a semicubical parabola, "van Heuraet's curve," as he labeled it. A few pages later, at the end of his work on evolutes, he derived, in terms of the parameters of his original hyperbola, the auxiliary curve whose quadrature van Heuraet's theorem states is equal to the rectification of that semicubical parabola.[85] He did not attempt to find corresponding auxiliary curves for his other evolutes, and it is doubtful that he would have pursued the case at hand if the form of the evolute had been other than a relative of the semicubical parabola, which quite naturally triggered the recall of van Heuraet's result, just recently received. The derivation is literally an aside, never mentioned again, not even in the *Horologium Oscillatorium*'s inventory of results.

However, if the evolute is treated as the locus of the centers of curvature of the hyperbola and if the entire set of derivations is written in differential notation so that its dependence on the tangent is evident, then the problem reduces to an expression that can be generalized into the fundamental theorem, as indeed van Heuraet's theorem itself can be reworked. Obviously, Huygens did not handle the derivations in this manner, so one can readily agree that his more geometric approach did not lead him to the fundamental theorem of the calculus. Nonetheless, it does not follow that his geometric approach was inconsequential and interfered with the rise of the calculus.

The growth of the calculus was a gradual development and was due in part to the change in perspective of the succeeding generation. Both Newton and Leibniz were relatively unschooled in Greek geometry, and much of their early work on the calculus gives the impression that they were searching for algorithms with which to master the world of line and form. Given their backgrounds, it is not surprising that they should emphasize the process by which a curve is generated over the curve itself, as indeed they did with respect to curvature, for example. Both mathematicians were greatly aided by the codification of the new analytic approach to geometry in the one textbook that they did study with care, none other than that propagandist tract for the Cartesians, van Schooten's *Geometria,* the second edition of which contained van Heuraet's theorem as well as many unsigned contributions from Huygens.

However, these practitioners of the new mathematics scarcely repudiated the past. Newton constantly attempted to refine his method of analysis by replacing fluxions with something more geometrically meaningful and thus acceptable, and his first biographer would recall:

Sir *Isaac Newton* has several times particularly recommended to me *Huygens*'s stile and manner. He thought him the most elegant of any mathematical writer of modern times, and the most just imitator of the ancients. Of their taste, and form of demonstration Sir *Isaac* always professed himself a great admirer: I have heard him even censure himself for not following them yet more closely than he did.[86]

Likewise, Leibniz can be found, again and again, voicing the old standards: "I claim also to be able to reduce always quadratures to the dimensions of curves, which I take to be simpler."[87]

The rise of the calculus was not a revolution *against* geometry but rather a slow evolution within the new geometry. It began with the algebraic representation of specific geometric problems according to the methods set down by Fermat and Descartes. As the methods became more familiar and standardized, underlying manipulative rules such as Hudde's "differentiation" rule and van Heuraet's transformation became apparent. Thus, the focus gradually shifted from specific problems to the determination of algorithms, and the geometric basis became less important. Evidence of this shift can be noted in the difference between ancient descriptions that classify according to the physical generation of the curve (planar, conic, mechanical) and modern ones that derive from operational procedures applied to a function (analytic, twice-differentiable, Fourier transform). The essential change that made possible a fully analytic calculus had more to do with the growth of the operator concept than with a repudiation of geometry per se.

Certainly, geometry retained its hold on the calculus well into the eighteenth century, if for no other reason than that early analysis was often stimulated by ideal physical problems that were expressed in geometric terms. In his classic study of the Leibnizian calculus for *Archive for History of Exact Sciences,* Bos advances a similar view. He describes infinitesimal analysis as an outgrowth of the Cartesian application of algebra to curves and does not see it separating from geometry until well into the new century. Indeed, part of his thesis is that the Leibnizian calculus succeeded in leaving behind the restrictions of its geometric foundations only when it shifted its emphasis from the differential, a dimensional measure, to the derivative, a dimensionless ratio. Yet as Bos cogently argues, the geometric phase, although eventually supplanted, was essential to the growth of the early calculus because it permitted a flexible approach to problems and led consequently to the elucidation of the dual concepts of independent variable and function that are prerequisites of a free-standing analysis.[88]

In sum, the geometry Huygens practiced was no mere foil to the rise of the calculus. The mathematics of the third quarter of the seventeenth century was a natural, even necessary, prelude to the growth of analysis. No longer ancient, it was not yet modern. It still clung to the old problems regarding specific curves and required the answer to be phrased in the proper pristine, synthetic fashion – everything in proportion, in every sense. Yet by incorporating infinitesimal techniques and analytic expressions, it handled current problems very successfully. The acknowledged master of this intermediate geometric analysis, as Newton proclaimed, was Christiaan Huygens. His application of evolutes to rectification rather than to curvature reveals his bond to the ancients, but his derivations of the evolutes themselves and of the isochronism of the cycloid point the way to the future.

Unfortunately, irrespective of its relationship to the rise of the calculus, Huygens's method of rectification by means of evolutes had a flaw, and an example reveals the limitation. In addition to the cycloid and the parabola, for which he derived his first two evolutes, there had been a third curve also founded in his physics that had contributed to the development of his theory. In *De Vi Centrifuga,* a man releases a ball while standing on the edge of a circular platform rotating in gravity-free space. Relative to the man's position, the released ball moves along the involute formed by unrolling a thread off the circular platform. However, although he could mechanically draw the involute of a circle, or of any other curve, Huygens could not necessarily mathematically derive its equation, which meant that he could never rectify the circle, or any other curve.[89] His method of rectification was always a two-stage process. Given a curve, its evolute was calculated, and the distance of that evolute from the original involute then determined the arc length of the *evolute.* It was the companion curve that was rectified, not the given curve. Although innumerable arc lengths could be measured in this way, given a specified curve, it could not be rectified by this method unless its involute was recognizable. The difficulty underlying this flaw in Huygens's method is common to many precalculus techniques involving the precursor of the integral and is eliminated by the antiderivative approach to integration made possible by the elusive fundamental theorem.

However, this weakness in the theory of evolutes as a method of rectification should not be allowed to obscure the fact that it was the first general technique for actually measuring the lengths of curves rather than merely transforming them. Using it, Huygens had found the arc lengths of a whole class of curves not known to have been previously rectified, namely the higher-order conics.[90] In addition, all the rectifications known to his contemporaries by other means could be subsumed under the same technique. The *Horologium Oscillatorium,* into which Huygens would pour his myriad discoveries, was one of the supreme mathematical treatises of the seventeenth century, as well as one of its principal physical tracts.

8

Diversions

Often responding to questions stated by others and exceedingly proud of his ability to find the answers, Huygens obviously desired the acclaim of his contemporaries. He constantly compared himself to the illustrious forefathers of science, undoubtedly searching for the recognition that was theirs. Yet the erratic way in which he communicated his results guaranteed a diffuse and diverse reception and ultimately undermined his place in history.

Certainly, the contemporary response to Huygens's theory of evolutes and its companion discoveries cannot be judged by the notice given to the *Horologium Oscillatorium,* which was not published until fourteen years after Huygens had created the mathematical technique, by which time most researchers had already learned of its major applications through more informal channels of communication. Indeed, many of the people with whom Huygens had scientific ties heard about the cycloidal clock and the isochronous path of its bob within a year of its creation.

Of course, van Schooten had been told about the new design almost at the moment of its birth, and a month later a similar description went to the Flemish mathematician Andreas Tacquet. However, although both letters contain a sketch of a pendulum hung between curved plates, neither identifies the curve as a cycloid.[1] Instead, the derivation was left as an unspoken challenge to the recipient.

Subsequent letters to Jean Chapelain, Pierre de Carcavy, Boulliau, and Wallis are even more vague, with no sketch and only a brief claim to have created an isochronous clock.[2] When pressed by Carcavy for more details, Huygens defends his reticence by saying that he is in the process of writing up his results for publication, and "I see more than ever by the latest writings of Wallis the inconveniences and disputes that can arise when discoveries of any consequence go from hand to hand before being published." He adds that the forthcoming work will also include "my method of measuring curved lines, which I do not believe anyone has heretofore conceived."[3]

Various manuscripts do substantiate that Huygens was revising and refining his derivations, presumably for eventual publication.[4] His Parisian friends erroneously assumed from his letter to Carcavy and from the remark made to Boulliau regarding a revised edition of the 1658 *Horologium* that the new work was almost completed and eagerly awaited the promised treatise. When a projected diplomatic trip by Huygens to Paris on behalf of his father was repeatedly postponed, Chapelain expectantly enthused in a letter that at least Huygens should have had time to finish his manuscript while waiting.[5] Indeed, Huygens replied that the treatise had been finished for quite a while, but that he had had no time to arrange for its printing before leaving![6] The comment to Chapelain was to be only the first of many excuses offered to his friend regarding the delay in publication, and when he finally departed for Paris in October 1660 his baggage did not contain a revised *Horologium*.

Once settled in Paris, he had little recourse but to pass the information "hand to hand" after all, for he could scarcely refuse to communicate his results when pressed in person. Foremost, the cognoscenti that gathered every Tuesday at the academy of Henri-Louis Habert de Montmor were told, although it is questionable whether Huygens also discussed the failure of his first attempt to determine longitude at sea by means of a cycloidal clock.[7] Certainly, by the end of November he was no longer concealing the particulars of the clock's design. For example, a letter to Leopold de' Medici not only includes a sketch of the curved plates but also specifies their cycloidal shape.[8] In fact, frequent references in his travel diary to the clockmaker Martinot and subsequent letters with Parisian correspondents make clear that Huygens was intent primarily on overseeing the dissemination of this marvelous invention.[9] By the time Huygens left Paris in the spring, his clock per se had instigated a sizable fuss, but its theoretical aspects had generated no comparable reaction, probably because Huygens had managed to avoid presenting the mathematics underlying the mechanism.

Yet it was not for his clock alone that Chapelain praises Huygens soon after his departure: "Monsieur Montmor has asked me a hundred times if I've had news of you and if I haven't received any letters from you. One thinks of you always in his Academy and

you have left an air of probity there that will linger for a long time."
Of course, Chapelain could not restrain himself from pushing for
the rapid publication of the new treatise on pendulums, even urging in the name of the "Public, all that you are capable of giving
it for its instruction and profit."[10]

The scene was repeated at Huygens's next port of call, London.
As in Paris, he was there ostensibly as his father's surrogate in a
diplomatic mission, but once again he found ample time to develop
his scientific contacts, and within four days of his arrival he was off
to one of the meetings regularly scheduled at Gresham College.[11]
Again, although his diary gives few particulars, later letters reveal
that he told many of his contacts about the cycloidal clock.[12]

How exhilarating this trip must have been for Huygens! Writing
to his younger brother while still in Paris, he marvels at how very
well he has been received on account of his clock and his *Systema
Saturnium*.[13] Moreover, in London he added to those two calling
cards the feat of correctly analyzing, for an august company that
included Wallis, Wren, Brouncker, and Sir Robert Moray, various "cases they proposed to me touching upon the impact of two
spheres."[14] In both cities, the men with whom he discussed nature
and art were busy laying the foundations of the twin pillars of in-
stitutional science: the Royal Society of London (chartered 1662)
and the Académie Royale des Sciences (1666). Given his impressive
display of talent during the trip of 1660–1, reinforced by another
trip in 1663–4, it seems inevitable that this facile Dutchman would
have been tendered membership by both national organizations.[15]

Indeed, to a large degree, Huygens's international stature during
the 1660s *was* the reaction to the *Horologium Oscillatorium*, albeit
premature by thirteen years. In 1663, while debating what kind of
permanent institutions would best reflect the glory of Louis XIV,
his minister, Jean-Baptiste Colbert, granted a largesse to certain
significant figures in the arts and sciences. Although most were,
to use Oldenburg's description, "Poets and Romancers," Huygens
was one of three "learned men" on the list who were destined for
the Académie Royale des Sciences.[16] Specifically, 1,200 livres were
to go to the man whose talents were summarized by the phrase
"great mathematician, inventor of the pendulum clock."[17]

Quite clearly, Colbert expected a return on his investment, namely
the dedication to Louis XIV of the new treatise whose publication
Chapelain and his scientific friends were still awaiting in 1663. Later

letters suggest that during his audience with Colbert, in which Huygens thanked the minister and king for their generosity, he himself announced his intention to dedicate a work to Louis. Yet two years later, even as Colbert was conferring upon him a second grant and was finalizing plans to bring him to the Académie, Huygens was not prepared to publish the new book, Chapelain's incessant reminders regarding protocol notwithstanding.[18]

Why did he wait to publish the *Horologium Oscillatorium* when he did possess most of its contents by 1660, if not the treatise in completed form? Certainly, his travels in 1660–1 and 1663–4 hindered any concentrated attempt to organize and improve his demonstrations. However, perhaps a more decisive cause of the delay in publication was not his travels but the travels of his clocks.

SEA TRIALS

In a letter to his father during negotiations for a French patent, Christiaan explains that on land a simple pendulum can be made to go as accurately as a cycloidal one and for less money. The simple-pendulum clock was cheaper for the obvious reason that the curved plates of the cycloidal pendulum, while allowing for unlimited amplitude of swing on a swaying ship, added to the expense and difficulty of the clock's construction. Thus, a nonmarine version of the cycloidal clock was of little practical value. Indeed, Huygens was so completely unconcerned about protecting his priority on land that only the opportune intercession of Carcavy extended the French patent to include the nonmarine cycloidal design, and the Dutch patent negotiated by Huygens himself contained no such clause.[19]

Although Huygens could have easily published an account of his theoretically perfect cycloidal clock and its application to the problem of the constant of gravitational acceleration, he chose instead to prove the clock's superiority over its simple-pendulum cousin as a practical instrument. Of course, the simple-pendulum clock had first achieved accuracy primarily through his own improvements in design, which he described in the *Horologium* of 1658, and his trips during the early 1660s had popularized those designs along with the more esoteric cycloidal clock. Yet, once discovered, the elegant mathematical ideal that informed the cycloidal clock's design forever prejudiced Huygens in its favor. Surely, mathematical superiority must be reflected in reality.

To compete in the real world, however, the clock would have to go to sea, where successful trials would prove the inherent advantage of the cycloidal design by solving the intractable problem of longitude. Not incidently, but by no means primarily (as some would claim), success at sea would also greatly increase the value of his patents and bring him further acclaim. Unfortunately, these interrelated goals proved frustratingly elusive, and in the meantime the revision of the *Horologium* was withheld from publication pending a definitive outcome.

While Huygens was in Paris in late 1660, the first long-distance trial was being conducted by his younger brother, who took a cycloidal clock with him on a voyage to Spain (on still another family diplomatic mission). Christiaan was not greatly perturbed by Lodewijk's glum report that pendulum clocks were useless on a tossing ship, because he had also received news that the same storm that had caused his brother such havoc had been severe enough to have wrecked five ships of the Dutch merchant marine. He was more concerned to learn how matters fared during reasonably temperate weather, rationalizing that in extreme cases the clocks could be stopped (for their protection) and reset at the next port of call.[20] His argument is somewhat specious, since the need for accurate positioning is most acute after a driving storm. Nonetheless, although the trial was a failure, Huygens remained optimistic that a clock specifically designed for ocean travel was feasible.[21]

Two years and many mechanical changes later, his enthusiasm was waning, in part because he was losing control over his invention. He was forced to become partners with Sir Alexander Bruce when the latter greatly improved the seaworthiness of his design and claimed thereby the right to patent the clock. Although Huygens reluctantly conceded him an equal share, Bruce demanded more, belittling Huygens's contribution as no more than an obvious synthesis of preexisting parts. Needless to say, to Huygens the theoretical aspects behind that synthesis were the raison d'être of the clock and Bruce's claims were insulting. Exacerbated by the regulation that a foreigner could not apply for an English patent, the dispute was mediated by Moray, with the result that Bruce shared the patent with the Royal Society, which not only acted as Huygens's surrogate but also became a partner in its own right.[22]

Moreover, the sea trials that would determine the value of that patent were in the hands of Huygens's new British partners, and all he could do was wait for their reports. In spite of Bruce's vaunted improvements, the maiden voyage (1662) of the marine clocks was a disaster, one clock crashing down from its mount and the other tossing uncontrollably until it stopped altogether.[23] No wonder Huygens was becoming depressed, having spent so much of his time in the preceding months in intense collaboration over the clocks only to send them off to their immediate destruction.

However, from the shattered remains of their originals, Bruce soon had two new clocks ready to accompany a fleet sailing to Lisbon under Captain Richard Holmes (1663), with moderately successful results.[24] This encouraging turn of events prompted the Royal Society to send the clocks with Holmes on a lengthy voyage that his fleet was to make to the South Atlantic. Upon his return in 1665, Holmes initially gave a glowing oral account, including the dramatic story of a decision, made when the fleet ran short of water, to entrust the determination of his position to the clocks rather than to the conflicting calculations of his pilots.[25] Hearing of this unqualified success, the astonished – and skeptical – Huygens tells Moray that the accuracy with which a small island was found was far better than he could ever have reasonably expected, which leads him to ask politely whether Holmes's word can be trusted.[26] Ascertaining the answer proved very difficult and time-consuming for Moray. His problem was compounded by the fact that Holmes was being shunted in and out of the Tower for unauthorized raids (against Dutch holdings!) that took place during the same voyage, which is not exactly a testimonial to Holmes's character. After many months, Moray finally found a pilot from Holmes's expedition who confirmed that the clocks had performed very well indeed, despite discrepancies that Moray's investigation had uncovered.[27]

Although it had been the most extensive and successful trial that the marine cycloidal clocks had undergone, the denouement of the episode was far from ideal. The account given orally by Holmes never became an official report, and Moray had to submit the formal patent application without its substantiating evidence.[28] However, Huygens continued to delay, hoping to append Holmes's account to his new treatise. At last, he too proceeded without the report, contenting himself with a brief summary of the excursion's findings, and even that synopsis was attached not to a revised *Horologium* but to an instruction pamphlet for pilots on how to use cycloidal-pendulum clocks at sea.[29] Variants of that summary, all drawn from a letter that Moray had sent to Huygens, had already appeared in the *Philosophical Transactions* and the *Journal des Savants* soon after the initial report by Holmes, so that the public record of the sea trial, although not containing a firsthand report, certainly leaves an impression of unqualified success.[30] Yet a revised *Horologium* still did not appear.

THE UNIVERSAL MEASURE

Unfortunately for Huygens's impatient friends, while waiting for vindication at sea, Huygens had encountered new snags on shore. For one, certain members of the Royal Society – particularly Robert Hooke – had begun raising questions about his original work in determining the constant of gravitational acceleration. The matter began felicitously enough when, still fascinated by Huygens's new isochronous clock, the Society decided to investigate whether the length of Huygens's cycloidal pendulum could become the basis for a universal unit of measure.[31] "You will perhaps reproach us, with justice, for the impatience with which we leap to make this experiment even before your treatises are published. Yet," Moray continues in his report of the session, "you will not reproach this impatience, considering the eagerness that we have to see your treatises."[32] As the new year began and before the Society could even procure clocks made to Huygens's design, Moray reports that four or five members had each been charged with separate experiments involving bobs of differing materials and weights, with variable results. He continues:

The Experiments that we have made concerning your curve [the cycloid] have succeeded so magnificently that everyone is well satisfied concerning it. Nonetheless, as to the precise exactitude of the motion of the Clock's pendulum I would be pleased to know if you are able to observe that it always is so equal that variations in the air do not affect it.[33]

And a week later, he asks Huygens to send a sample length in Rhenish inches for the experimenters (a clear indication of the need for a universal measure), adding:

I believe I told you in my preceding that the balls of lead of differing weight did not make equal swings in the same time, except when the swings were small, according to what we have tested, although it seems that you have not found any difference in differing materials or weights.[34]

The question Moray was too polite to ask directly is obvious: Are you sure?

Whatever the "Experiments" regarding the cycloid might have been, the Society was not at all concerned with the theoretical, mathematical background of the clocks (except Brouncker, who began a series of attempts to prove the isochronism of the cycloid). Rather, the members were interested in the accuracy of the physical model, on land as much as at sea. However, as was typical for the Royal Society, they soon became distracted and abandoned their study without coming to a definitive conclusion, except that "Dr WREN was desired to think of an easy way for a universal measure, different from that of a pendulum."[35]

In the meantime, not to be outdone by a foreigner, Hooke was busy concocting his own machine for measuring the constant of gravitational acceleration.[36] While the rest of the Society wandered away to other concerns, Hooke continued to raise objections to Huygens's clocks and to draw attention to his own competing devices. Thus, it was not without prejudice that Hooke undertook exacting experiments before the Society in December 1664 in order to test Huygens's latest result, the mathematically derived formula for the center of oscillation of a sphere suspended by a weightless cord (Hooke used a hair).[37] The first round yielded values "very near" the predicted lengths, but when Hooke introduced longer cords (hairs) in the second attempt the values deviated by as much as 15 percent, "which was thought too considerable a difference for a standard. It was thereupon concluded that there must be either some mistake in the rule, or some fault in the experiment."[38] Ignoring any possible experimental failure (such as the elasticity of the long hair), Hooke tells Robert Boyle that Huygens's rule "is now much doubted of."[39]

In addition, Huygens fueled Hooke's attack by reporting to Moray that, irrespective of the initial positions of their bobs, two clocks would become synchronized if they were in close proximity.[40] At first, Huygens had even viewed the synchrony as a very positive result, for he reasoned that unavoidable minor anomalies in the mechanical going of one clock would be smoothed out by its sympathetic response to its partner at sea: "This discovery has pleased me no little, being at the same time a beautiful proof of the exactness of these clocks, since it takes so little effort to maintain them in perpetual accord."[41]

Even as Moray was replying with doubts and suggestions, Huygens was composing a retraction of his erroneous explanation of the synchroneity. The sympathetic motion arose not from subtle variations in the air pressure caused by the swinging bobs, but rather from slight movements of the tandem mounting used to support paired marine clocks.[42] Since Moray had dutifully communicated Huygens's reports to the Society, new objections arose: "Occasion was taken here by some of the members to doubt the exactness of the motion of these watches at sea, since so slight and almost insensible motion was able to cause an alteration in their going."[43]

This new obstacle, coupled with Hooke's ongoing objections, cooled enthusiasm considerably, overriding the momentary excitement created by Holmes's positive report. In addition, much to his embarrassment, Huygens's eager friends in Paris had published the similar hasty announcement and erroneous explanation of the phenomenon that he had sent to his father, who was currently in the French city.[44]

Thus, Huygens was placed on the defensive. In fact, Chapelain cleverly used the attacks to pressure Huygens once more regarding the publication of the promised revision of the *Horologium:* "Someone has written from up your way, from envy no doubt, that you have even found some defects in your Invention that prevent you from publishing it."[45] Of course, Huygens indignantly denied the charge and retorted that nothing compared to the exactness of his clocks. As to the treatise, well, it was done but he had many things to add and no time in which to properly address them before leaving to take up residency with the Académie Royale.[46] Indeed, six years after the first time he invoked that claim, he would again embark for Paris with the *Horologium Oscillatorium* incomplete.[47]

THE COMPOUND PENDULUM

Nevertheless, it is true that he had new discoveries to add. His work of 1659 had been inspired by Mersenne, and even in the midst of his study of the ideal simple pendulum he had begun a tentative analysis of the compound pendulum, thus returning to a related problem in which Mersenne had tried to interest the young Huygens so many years before.[48] The attempt of 1659 concentrated on incorporating into his mathematics the weight of the cord (often in reality a solid bar with a fine chain at the top for flexing around the cycloidal plates), by treating it first as another point-mass and then as a series of point-masses spread along its length.[49] The analysis was abandoned, however, in favor of work on generalizing evolutes to the conics.[50]

Only after his return from a year in France and England did Huygens begin a study that treated compound pendulums, ones whose cords and bobs have weight. Had his friends, even in the flush of their enthusiasm over his invention, already expressed their concerns for the physical manifestation of his theory? Or was it merely his first opportunity to pick up the thread of research interrupted by conflicting obligations? For he did begin where he left off, transferring the earlier sketch of 1659 into the beginning of still another compact treatise. Even before Moray could inform him of the Society's formal decision to investigate the pendulum as a universal measure, he had finished his analysis of the effect of weight distributed along the cord.[51] A byproduct of this study was the introduction to his clocks of the *curseur* weight, which when placed at the center of oscillation of the pendulum had no effect but when placed slightly above or below that point altered the period, thus allowing for fine adjustments in the clock's timekeeping.[52]

After this brief period of study at the end of 1661, Huygens once again abandoned his analysis of the compound pendulum, just as the Royal Society abandoned its interest in his clocks. Were both he and the Society waiting for the news from sea? Seeming to wax and wane in parallel with the Society's interest, Huygens's research on compound pendulums did not resume until after his return from a second mission to France and England (1663-4). From the new round of research came the test case regarding the center of oscillation of a sphere that Hooke undertook for the Society in the December 1664 meetings.[53]

The sphere was only one of a great variety of geometric entities for which Huygens derived the center of oscillation. Triangles, semicircles, cylindrical wedges (*ungulae*) – all those objects that Mersenne had written about, requesting his analysis, so many years before – were finally treated. Appropriately, his research for the *Horologium Oscillatorium* ended where it began, with a problem posed by Mersenne. When combined with the treatise of 1661, the study comprised the most complete analysis of the compound pendulum to date. Part 4 of the *Horologium Oscillatorium* was thus finished, and there would be no further additions of major importance.

The move to Paris, however, afforded a new round of excuses for delaying publication similar in scope to those already encountered. Interfering obligations to his father were replaced by the demands of king and academy. As usual, the interests of others influenced his research choices, in some cases causing delays posterity can hardly begrudge. For the Royal Society he codified his work on collisions of hard bodies; for the Académie Royale he read his first draft of a discourse on the cause of gravity.[54] Concomitantly, access to royal ships allowed for further sea trials of his marine clocks.[55] Perhaps the results would be more decisive once he was in charge, for, he confessed to Moray, he continued to strive for the "perfect clock."[56] Still further delays arose when, seriously ill and afraid he was dying, he returned home for a year (1670–1).[57]

Suddenly aware of his mortality, he also learned the limits of Colbert's patience. Thus, restored to health and to the French court, Huygens finally published his masterpiece with its promised dedication to his patron, Louis XIV, who at that moment (1673) was occupied with invading Huygens's homeland. The price for delay was paid.[58]

Once published, the *Horologium Oscillatorium* generated very little reaction. In all probability, because most of the major results had been known for a long time, very few people actually applied themselves to the exacting mathematical derivations that underpinned those results. Indeed, having learned the conclusions years before, many mathematicians adept enough to understand the derivations had constructed proofs of their own in the meantime. In particular, Brouncker and Pardies had produced completely different proofs of the isochronism of the cycloid and, warned of Huygens's impending publication, rushed their proofs into print in order to protect their priorities regarding methods of solution.[59] Even Newton seems to have heard of this famous discovery somehow, for his papers also include a derivation of the isochronism of the cycloid, likewise starting with the assumption that the cycloid is the necessary curve.[60]

Thus, even the mathematics, when finally made public, was not as influential as it might have been had Huygens published his discoveries a decade earlier. Upon reading his copy "with great satisfaction," Newton remarks to Oldenburg that he finds it "full of very subtile and usefull speculations very worthy of ye Author." Yet by that time he had already duplicated much of its content, and he ends his brief discussion of the book with his offer to send his easier method for rectifying curves. Above all, Newton expresses his eagerness to see the treatise on gravity that Huygens promised to publish soon.[61] Having withheld his proofs of the theorems concerning the conical pendulum and centrifugal force, Huygens states in Part 5 of the *Horologium Oscillatorium* that he plans to issue a complete treatise on gravity soon.[62] Presumably the projected treatise would have joined the discourse on gravity presented to the Académie with the withheld derivations. Since Newton had once independently derived the conical-pendulum formula and had already made his first attempt to analyze the moon's circular motion around the earth, his eagerness to see what the great master knew was understandable, and unrewarded.[63]

For Newton, already set upon his own course by the time it was published, the *Horologium Oscillatorium* certainly had little to offer as far as content was concerned. Nonetheless, he recognized in it a supreme model of how to apply mathematics to physics and consciously attempted to follow it in the *Principia*. In 1691, when Richard Bentley encountered difficulties in understanding the *Principia,* Newton prepared a list of background readings for him that included the works of van Schooten and Barrow: "These are sufficient for understanding my book: but if you can procure Hugenius's *Horologium oscillatorium,* the perusal of that will make you much more ready."[64]

For the other giant of the new mathematics, the receipt of the *Horologium Oscillatorium,* direct from the hands of the author, symbolized how much he still had to learn about a field in which he was both genius and novice. When Leibniz received his copy while on a visit to Paris, it was accompanied by encouragement tempered with advice on how to correct his deficiencies.[65] Yet even as he read to learn, he interpreted Huygens's work from his own viewpoint and thus, among other things, envisioned "osculating circles." Years later he would tell Huygens that the *Horologium Oscillatorium* was "one of the greatest incentives to whatever advance I have made since in these kinds of knowledge."[66]

With others, granted less creative insight than Newton and Leibniz, Huygens's mathematics was absorbed more completely whole. The prime example is Leibniz's lesser known compatriot and fellow traveler to Paris, Ehrenfried Walther von Tschirnhaus, who studied the theory of evolutes and assimilated it into his own work on generalizing Descartes's ovals.[67] Of course, it could be argued that his absorption of evolutes was a little too whole while not quite complete, for he later submitted a paper to the Académie Royale based on a new application of evolutes communicated to him in private by Huygens, only to have a committee from the Académie reject his proofs as grossly erroneous.[68]

Outright negative reactions to the *Horologium Oscillatorium* were primarily British in origin and involved secondary issues. Wallis objected to Huygens's history of rectification, with its emphasis on van Heuraet's contribution over that of Neil and consequently of his own. Hooke continued his complaints regarding priority of inventions, including now the conical pendulum.[69] To the south, the Italians, particularly Leopold de' Medici and Viviani, revived their protests regarding the slight to Galileo's reputation.

Even more persistent than any of these claims, however, were the efforts of the Abbé de Catelan in France to discredit Huygens's section on the compound pendulum. Many of the great Continental mathematicians who counted themselves as Huygens's followers were to get caught up in the futile attempt to silence him. The decidedly uneven debate generated some important consequences, although not from Catelan.[70]

FURTHER DISCOVERIES

Generating relatively little response from others, evolutes continued to function within Huygens's own work as a powerful mathematical tool. As so often before, a question from someone else stimulated him to return to the calculation of evolutes for specified curves. Pierre de Vaumesle asked him to find the evolute of an epicycloid, and in response Huygens demonstrated for the Académie that the epicycloids form a closed family of evolutes in the same way that the cycloids have cycloids as evolutes.[71]

In addition, Huygens persisted in his attempts to create accurate, seaworthy clocks, and the theory of evolutes provided the conceptual basis for the newer designs, just as it had with the first two. Not long after the publication of the *Horologium Oscillatorium,* in a return to his analysis of fall along an inverted cycloid, Huygens noticed that the effective weight of the bob is proportional to the length of the arc of the cycloid from the vertex to the bob's position. In modern terms, the restoring force is proportional to the displacement. He immediately extended this result regarding the requirement of isochronism to other oscillating bodies, including a vibrating cord, and a year or so later he applied it to springs. In the latter case, his primary technique was the same approach that had been so effective in dealing with centrifugal force, namely the balancing of counteracting tensions under the assumption that, if their effects are the same, the forces can be treated as equivalent no matter what their ultimate causes.[72]

A great variety of clocks arose from the confluence of this new insight into isochronous oscillating bodies and Huygens's earlier discoveries of isochronous curves. He designed a spring-driven pocket watch and thereby precipitated still another quarrel with Hooke. He also constructed a large marine version but soon learned that fluctuations in springs were no easier to control than perturbations of pendulums.[73]

In the original marine version of the basic cycloidal-pendulum design, Huygens had actually mounted the pendulum with two cords, attached to the clock fore and aft between very wide cycloidal plates, so as to lessen the effect in that dimension of a tossing ship (Fig. 8.1).[74] In 1683, in still another variation on his basic cycloidal design, he created an intricate pendulum regulator that consisted of a ring attached to the clock by three cords in such a manner that it oscillated approximately isochronously. Then he added curved plates to each point of the triangular mounting. Naturally, the plates were evolutes of the appropriate curve that would make the motion of the regulator truly isochronous, and the mathematics of their derivation is particularly impressive, because the system was three-dimensional.[75]

Turning to the balance clocks of Galileo's time, Huygens applied the knowledge gained during his work for *De Vi Centrifuga* to guarantee the isochronism of that oscillating system. One fascinating design shows a balance mounted on a vertical axle with grooved arms shaped to form a parabola (Fig. 8.2). Inside the grooves metal balls were to rise and fall in response to the accidental variation in the oscillations of the balance.[76] Thus, Huygens wedded the isochronism of the paraboloid to the balance clock.

For a balance with a horizontal axle, he revived the device, shown in his early design for the conical pendulum, of stacking the links

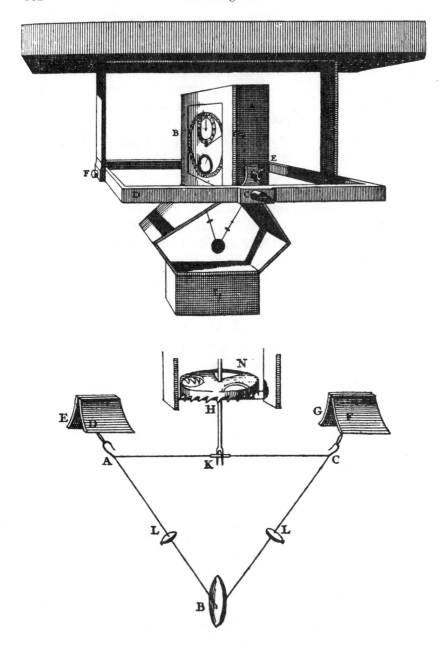

Figure 8.1. Marine cycloidal clock with close-up of double suspension.

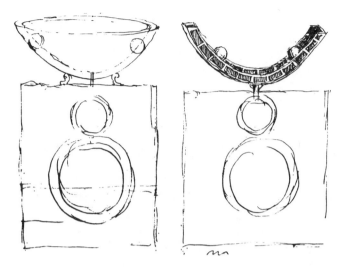

Figure 8.2. Development of the parabolic balance regulator.

of a chain in order to maintain proper oscillation, one pile of links for each arm of the balance rod (Fig. 8.3).[77] Then he hit upon a system in which a pair of curved plates with a heavy weight hung between them was rigidly attached to the horizontal axle of the balance. The weight was not intended to swing like a pendulum but to hang vertically from the rotating system. As the balance oscillated, the bob traced out a parabola. Amazingly, the resulting system, as well as another variant developed later, was theoretically isochronous. In creating this system, Huygens finally found a horological application for the third physical curve that had inspired his original research years before, because the curved plates on this clock were involutes of a circle.[78] Evolutes, when combined with his knowledge of oscillating bodies, seemed to guarantee isochronism wherever he turned.

CAUSTICS

However, the physical application of evolutes that influenced other mathematicians, although it was eventually forgotten and rediscovered by still others, was his use of the theory to determine the form that the wave front of light would take upon reflection or

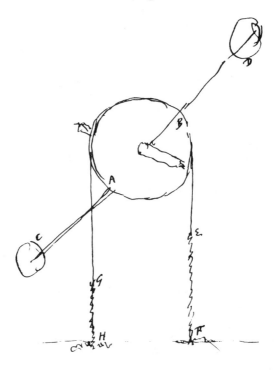

Figure 8.3. Balance regulator with chains.

refraction.[79] In 1677, while convalescing at home, he returned to his ongoing study of dioptrics. After attempting to handle Fermat's principle of least time and Descartes's ovals of refraction (now called aplanatic surfaces), in which light from a point source is refocused to a single point, Huygens turned to more general situations in which light does not reconverge after refraction. In particular, he considered the case in which a spherical wave front passes through a planar interface and the case in which parallel light (a linear wave front) is refracted through a surface with a circular cross section (most likely he was thinking of a sphere, but a cylinder qualifies). From these specific derivations, he leaped immediately to a generalization, recording a passage in the same rigorous, somewhat didactic style of his other compact treatises:

The curve commonly tangent to all the particular waves will be the propagation of the principal wave in glass. Therefore, the straight lines that cut this common tangent curve at right angles will be the refracted rays.

These, however, are given from another source. Therefore, they themselves cut that curve at right angles. Therefore, the curve arises from the
evolution of the other curve, which is the common tangent of these rays.

It suffices to know that the waves are propagated in the glass along
these straight lines. But since the straight lines ought to cut the wave at
right angles, it could appear astonishing how the lines not tending to one
center can always cut the wave at right angles. But this now is explained
by evolution.[80]

In other words, he claims that even if the rays do not converge after refraction the wave front is not lost, but can be determined by
the theory of evolutes, because the refracted wave front is the involute of the curve that has the refracted rays as its tangents. An
analysis of Iceland spar follows, culminating in his declaration of
"Eureka" at discovering the explanation of double refraction by
means of evolutes and wave fronts.[81]

Two years later, having returned to Paris, he presented many of
these findings to the Académie Royale but, true to his fashion, did
not publish them until 1690. Only with the *Traité de la lumière* did
Huygens fully develop the idea boldly stated in the passage of 1677.
Chapter 6 of that work begins with his presentation of the Cartesian ovals and his generalizations regarding like surfaces. After
an exhaustive analysis of the cases in which light is refracted or
reflected to a point, Huygens continues:

Having presented the discovery of the curved lines that serve for the perfect convergence of the rays, there remains to explain a notable thing concerning the uncoordinated refraction of spherical, planar, and other surfaces, which if ignored could cause some doubt concerning what we have
said many times that rays of light are straight lines that intersect the waves
that propagate with them at right angles. . . . what can the waves of light
be in this transparent body, which are cut at right angles by the converging rays? . . . And what will these waves become after the said rays begin
to intersect one another?[82]

He answers his own question by meticulously deriving for the
case of the sphere the transformed wave front *EK* (Fig. 8.4) "made
by the Evolution of another curve *ENC*, which touches all the rays
HL, *GM*, *FO*, etc., that are the refractions of the parallel rays; by
imagining a thread laid over the convexity *ENC*, which by unwinding describes at its end *E* the said curve *EK*."[83] In modern terms,
ENC is the envelope of the refracted waves and is called the caustic; in Huygens's terms, *ENC* is the evolute and *EK* is its involute.

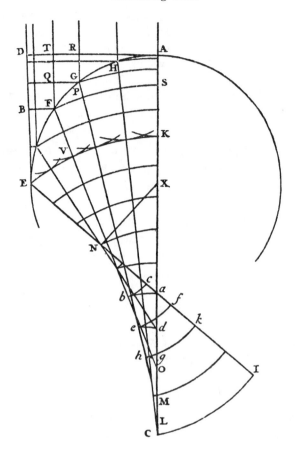

Figure 8.4. The transformed wave front.

Huygens concludes with his derivation of the wave front formed by the reflection of sunlight in a spherical mirror. In this case, the caustic (evolute) is his familiar friend, the epicycloid.

Although Huygens had derived the caustics of refraction (diacaustics) and reflection (catacaustics) only for the sphere, his method was completely general, and he obviously intended to provide proofs by example. The succeeding generation of mathematicians, particularly the Bernoulli brothers, would finish his work for him.[84] In the meantime, as anyone who has looked at the *Traité* can attest, Huygens concentrated on the mathematics of double refraction in Iceland spar.

These later applications of evolutes reflect the approach to nature evident in the original derivations of 1659. A physical problem is reduced to a few choice parameters capable of mathematical expression and hence manipulation, and the results of the mathematical derivations are presumed to say something valid about the real world. Is it the cycloid that is isochronous or the clock whose bob follows its path? The terminology can be loose because the pairing of mathematical entity and physical object is so precise. A complete mathematical description suffices as the solution of a physical problem. The wave theory of light must be correct because the mathematics of waves explains the double refraction of Iceland spar. Conversely, the vortices of Descartes must be rejected because Newton has shown them to be mathematically unsound.

Moreover, the physical should be made to match the mathematical as closely as possible. For example, Huygens's clocks tended to have relatively lightweight bobs and cords, thus approximating the ideal weightless situation, in contrast to the British long clock, which was developed soon after by empirical methods and was massive in every sense. In Huygens's world, an isochronous clock must have a regulator, such as a pendulum, that moves according to the mathematics of perfect oscillating motion. The curved plates were dispensable when they were the result of experimentation, but they became essential once they were the product of mathematical analysis. Whatever objections might be advanced regarding the actual value of cycloidal plates, such as the negative effects of the added friction, are irrelevant to this view. Physical aberration should be met by additional derivations, not by the abandonment of the design proved by mathematics.

If Huygens's work is an accurate exemplar, the mathematical phase of the Scientific Revolution consisted of the destruction of the boundaries between mathematics and the physical world: a new applied mathematics for a new mathematical physics. Mathematics became more than a model of correct reasoning, as it had been for Descartes and many other philosophers of nature. If God is a geometer, as Galileo declared, then all nature is ruled by mathematical principles waiting to be discovered. It was this attitude, more than any particular method such as the calculus, that the succeeding generations inherited from researchers like Huygens.

In sum, Huygens's theory of evolutes was quickly overridden by the calculus. The work done independently by Newton and Leibniz ensured that the essentials would still exist without Huygens's approach, albeit in a different form and for different reasons. The concept and terminology of the paired curves, evolute and involute, continued in the literature long after the theory ceased to function independently and can still be found in textbooks on differential geometry. Nonetheless, for the most part, Huygens's theory of evolutes died early, of neglect, and only the caustic saved it from immediate oblivion.

Yet broadly interpreted as an example of the application of mathematics to a physical problem, the theory of evolutes was hardly a failure. Huygens's work in the *Horologium Oscillatorium* and the *Traité de la lumière* provided a model for future research in mathematical physics. So rapidly and firmly engrained would this approach to nature become that Johann Bernoulli would write Leibniz in 1710,

He who undertakes to write Physics without understanding mathematics truly deals with trifles.[85]

9

Conclusion

What, finally, can be said about the man and his method and about
their significance? The pattern by which he worked is clearly evident. He was usually spurred to action by a problem posed by another, particularly if solving that problem carried prestige. Once
undertaken, it was usually pursued systematically and relatively rapidly to its solution, at which point the salient results were summarized in a short treatise. Because of his thoroughness, Huygens's
results often held much broader significance than the original problems from which they stemmed, but they were seldom linked together into a unified theory or applied to new areas, unless resparked
by other researchers.

Examples of motivation from the outside abound in Huygens's
work. Although his stature with Mersenne had been determined in
part by his proof that the catenary was not a parabola, he never
returned to the hanging chain to find its true shape until challenged
by Jakob Bernoulli.[1] Likewise, he never pursued an inconsistency
that he had noted in Bernoulli's analysis of inflection points until
he was questioned by l'Hospital.[2] In still another case, having come
to a different conclusion than Bernoulli regarding the sail curve,
he asked l'Hospital if he believed Bernoulli's results, concluding,
"Your authority would set me to repeat the examination."[3] Otherwise, it must be presumed, he would not bother to redo his calculations, since there was no guarantee that the effort would "pay."

Obviously, someone who depended so heavily upon outside stimuli could hardly adhere to a conscious research program. Huygens's
desire to be in the forefront of any area in which he had already
established a reputation meant that he tended to concentrate on related developments and to react to territorial intrusions. Although
his many priority squabbles should not be overemphasized, given
the fact that in all cases he remained in scholarly dialogue with his

opponents, it nonetheless remains true that questions of priority were as likely as questions of substance to promote research. Under circumstances like these, it is hardly surprising that metaphysical biases, although they undoubtedly could influence his responsiveness and shape his answers, could scarcely provide his goals.

The most telling attribute of Huygens's method of research is his consistent appeal to mathematics. Unlike Galileo, who invoked whatever argument would persuade his readers of the validity of his conclusion, be it analogy or experiment or mathematics, Huygens had only one approach to nature. Of course, he observed and experimented, especially when at the Académie Royale, but his greatest successes were those achieved with a pen, by a man possessed of a great logical facility, a born mathematician. Even his explanation of Saturn's ring owed more to deduction than to observation.[4] Likewise, his determination of the constant of gravitational acceleration had little to do with experimentation. Although his notebooks reveal many calculations of possible gear combinations and other necessary values, they disclose no table of repeated trials as might be found in Galileo's papers under similar circumstances.

Indeed, the choicest irony – or the profoundest significance – is to be had in the different approaches taken to the same problem by Huygens and the Royal Society's representative, Hooke. While Huygens mathematically derives the center of oscillation of his system, Hooke suspends heavy iron balls from hairs. While Huygens is completing his study of the compound pendulum, Hooke is redoing Riccioli's imperfect experiment.[5] Remember, the Royal Society never substantiated Huygens's values nor conceded to his derivations. The parallel streams of mathematics and experimentation did not converge in a flood of undeniable truth. Yet it was Huygens the mathematician, not Hooke the experimenter, who produced the first accurate measurement of a physical constant and thus provided Koyré with his paean to reason.

Huygens was fully cognizant that he could not ignore experience, but he was too aware that observation is biased and imprecise to grant it priority. It was enough, he would say more than once, that the experiments do not contradict the mathematics. Certainly, his treatment of Mersenne's experiment is a perfect example of this attitude. He trusted mathematics, and he would tell Moray, when the Royal Society had trouble confirming his results, that any anomalies were merely the effect of air resistance.[6] He was distressed that the Royal Society could not reproduce his values, but he was not deterred in his approach nor in his belief that he was correct.

Of course, Galileo quite often erroneously waved aside objections with allusions to unaccounted physical variables. Huygens's strength was his constancy. If objections persisted, he would try to incorporate the new variable into his formulas. Thus, he analyzed the mass of the bob and the mass of the cord and tried to subject air resistance to the same mathematical approach.[7] Every problem ultimately had a mathematical solution.

Naturally, this attitude had as much to do with his abilities as with any metaphysical commitment. He was playing to his strength. If Bruce belittled the theory incarnate in his clock, he would bind it in still more theory and repossess the prize. If Hooke objected to the omission of physical parameters, he would absorb them into the mathematics and remove them from Hooke's dominion of trial and error. In sum, bludgeon the enemy with the tool it did not possess, could not possess – mathematics.

Yet Huygens was capable of appreciating the talent of the enemy, even of that nemesis who claimed to have invented a better version of every instrument that Huygens had ever designed. When he first learned of Hooke's *Micrographia,* Huygens told his father that the author "does not understand geometry" and "renders himself ridiculous by his boasting," but once he had read the masterpiece he was unstinting in his praise, telling Moray:

For the rest, I am delighted to possess at last the said book of Mr. Hooke, which I had not imagined to be a volume of such importance. Certainly it is a very beautiful work and it is so curious that it was not printed for such a long time. I take such great pleasure rummaging through it that I have only reluctantly torn myself away in order to write you these lines. No one could produce more exact observations of this kind, nor better-made figures which must assuredly have cost him a great deal of unbelievable pain in order to plan how to execute the engraving so well.[8]

Hooke was far less generous in his attitude toward the *Horologium Oscillatorium.* How could he be otherwise? The mathematician could understand and thereby acknowledge observational skills; the experimenter was incapable of reciprocating.

Even among mathematicians of far greater talent than Hooke's, Huygens was the undisputed master of geometry, as Newton's assessment testifies. Certainly, the derivations reproduced in the first half of this book are evidence of his insight into the realm of line and curve. Even in the crudest sketches, the proper theorem from the appropriate, usually ancient source is effortlessly invoked. Of course, the grand exemplar of this amazing familiarity is his ready recognition of the cycloid as the curve that satisfied his geometric restrictions. Yet the mundane manipulation of proportions is equal evidence of his skill in making figures on paper yield meaning out of confusion.

Beyond the easy command of figures, however, lies the other aspect of geometry that Huygens had mastered, if indeed he had not been born with. The axiomatic structure of geometry was and still is the model for logical rigor. Its tightly argued proofs guarantee truth, or at least give the semblance of such assurance. However, not everyone is suited to its unyielding discipline, for it requires a highly analytical mind for whom order is paramount.

The many brief treatises scattered throughout the *Oeuvres complètes* testify to Huygens's predilection for mathematical rigor. They are not finished drafts for projected publication, although the editors of a posthumous collection of his works used them as such.[9] Rather, they are narrowly focused but complete analyses of specific problems. Usually, they consist of a set of theorems introduced by a paragraph that specifies certain physical presumptions upon which the theorems are based, such as Toricelli's principle or the equivalence of centrifugal force and gravity. The final stage of Huygens's process of discovery was this axiomatization of the results. The study was not complete until rigorous analysis had assured proximate certainty.

That proximate certainty applied as much to the physics involved as to the mathematics per se. If the theorems of his treatise formed a coherent whole and supplemented, or at least did not contradict, known results, then the assumptions on which those theorems were founded had a greater likelihood of being correct. Not that Huygens ascribed to physics the absoluteness of mathematics; even in the best of cases it was a realm of probable truths.

Nonetheless, to Huygens nature was fundamentally a geometric realm. In the *Cosmotheoros* he argues that people on other worlds would still develop Euclidean geometry because the same mathematical principles abide throughout the universe.[10] In other words, mathematics is not an abstract construct of our earthly minds but informs nature. As Dijksterhuis has pointed out, Huygens did not waste many words explaining why he felt that the Aristotelean division between the real and the mathematical was invalid.[11] He simply saw the physical world with the eyes of a geometer and wrote on the draft of *De Vi Centrifuga:*

Whatever you will have supposed not impossible either concerning gravity or motion or any other matter, if thence you prove something concerning the magnitude of a line, surface, or body, it will be true, as for instance, Archimedes on the quadrature of the parabola, where the tendency of heavy objects has been assumed to act through parallel lines.[12]

Huygens adopted Archimedes' approach to statics, already well developed by Stevin and others, and applied it to Galileo's science of motion, especially as extended by Mersenne, thereby developing an effective kinematics that moved well beyond his predecessors.

His grand derivation of the isochronism of the cycloid is a perfect example of his attitude in action. Free fall is nothing but a parabola; speed is finessed into a line; the instantaneous events of a world in flux are summed into areas. Every physical parameter is either absorbed into the geometry or eliminated by the use of proportions.

Yet even as he transferred his physics into geometric space, Huygens bound his geometry to the real world. Obviously, in the purely mathematical realm still dominated by classical problems, such as the determination of pi, he functioned in the abstract; but when the inspiration came from physical problems, the mathematics could not stray too far away. Referring to Leibniz's desire to extend the calculus, he tells l'Hospital: "I have not meditated a great deal on these matters, being always more satisfied in searching for the usefulness of Geometry in physics and mechanics."[13] To Leibniz he is even more emphatic:

You believe, it appears, that it would not be extremely difficult to achieve completely a Science of Lines and Numbers. So far, in that I am not of your opinion, nor even that it would be desirable that there remain nothing more to discover in Geometry. But this study ought not to prevent us from working on physics, for which I believe we know enough and more of Geometry than is necessary; but one must reason systematically with experiments, and amass from them data, approximately following Bacon's plan.[14]

L'Hospital's extravagant but truthful reply is to praise Huygens's unique abilities: "I agree that geometry is only a mental game unless applied to physics and mechanical inventions, but one who can succeed at it is rare and it takes entire centuries to produce a Huygens."[15]

Huygens's determination that mathematics not outrun physics made it difficult for anyone to be his intellectual heir, for what does a disciple do but extend and generalize the master's results. How does one follow a problem solver? Leibniz's answer was to go his own way, all the while tactfully granting his revered mentor his say. Constantly complaining about the penchant of the Bernoulli brothers to generalize, Huygens tells Leibniz at one point:

I would never amuse myself with the different kinds of catenaries that Johann Bernoulli proposes to achieve as before or urge on further this speculation. There are certain curves that nature often presents to our view, and which it itself describes, so to speak, those I judge worthy of consideration and they ordinarily contain several remarkable properties, as one sees in the Circle, Conic Sections, Cycloid, the basic Paraboloids, and the *Catenary*. But to invent from them new ones only in order to exercise his geometry, without anticipating there any other utility, it seems to me that it is *difficulties dealing with trifles* and I hold the same opinion regarding all problems touching upon numbers. *We play with pebbles* [the original meaning of *calculi*], *in order that we may grind subtleties into superfluities,* Seneca says somewhere in speaking of certain frivolous disputes of Greek philosophers.[16]

Leibniz's reply, diplomatically worded as usual, is an oblique defense of his own work as much as of Bernoulli's:

Regarding the curves of Mr. Bernoulli, you are right, Sir, to not approve if one amuses oneself researching curves invented for pleasure. I would however add a restriction: except if it can serve to perfect the act of discovery. That is why I do not disapprove of persons who have such leisure and inclination, and above all of young men, applying themselves to it. And by the same token I would also not discourage those who study numbers.[17]

Leibniz's vision of the algorithmic nature of modern mathematics won ascendancy, of course.

However, Huygens won out temporarily at least. The succeeding generation of Continental mathematicians expended most of its energy on interfacing physics and mathematics. In no area is Huygens's influence more noticeable than in the study of physical curves and related subjects that are today classified as rational mechanics. Ancient geometers had already introduced the center of gravity and mechanical curves into mathematics, but it was Huygens's successful analysis of the cycloid that prompted a series of studies regarding curves that satisfy a specified physical property, curves that include the caustic, the brachistochrone (the curve of fastest descent), the sail curve, and the catenary. The cycloid is a mechanical curve in the ancient sense of having its locus defined by a mechanical motion (the rolling of its generating circle), but it is also mechanical in a new sense, because it is the path of uniform timekeeping by a mechanical device. "Huygens's isochrone" Jakob Bernoulli called that path when he discovered that the brachistochrone is the same curve. Although he might wish to deny his paternity, Huygens's work was also the inspiration for more general analyses such as those by Jakob Bernoulli on Torricelli's principle, both Bernoullis on caustics, and Philippe de la Hire on epicycloids.

There remains the paradox that one of the supreme mathematicians of the seventeenth century was unable to accept one aspect of mathematics considered essential today, the generalization of method into algorithmic structure. In his defense, one can cite the fact that much of the speculative mathematics being produced toward the end of his life (the 1680s and 1690s) was not very firmly grounded. Indeed, the mathematicians of the early nineteenth century would spend most of their time axiomatizing and systematizing all that occurred in the century in between. In the meantime, creativity required a leap of faith, a willingness to follow the idea while leaving the details for later. The man who had to axiomatize his results before considering them valid could not accept this approach:

Your meditation on tangents by their foci appears to me to be very profound. However, it supposes certain things that cannot be admitted as evidence. Although some such reasonings can sometimes serve to invent, one needs in the end other means for more exact demonstration.[18]

Moreover, Huygens's strong sense of individual worth would never have let him be a slave to another's ideas, and a system of calculation, unlike the loose matrix of techniques that constituted seventeenth-century infinitesimal analysis, does render the user subservient. The clever gives way to the drone. Thus, Leibniz confesses to Huygens precisely the wrong goal:

What I love best about the calculus is that it gives us the same advantage over the ancients in the Geometry of Archimedes as Viète and Descartes have given us in the Geometry of Euclid or Apollonius, by freeing us from working with our imagination.[19]

The master of both the Geometry of Archimedes and the Geometry of Euclid and Apollonius was not so very ready to concede his imagination to Leibniz, particularly when he felt he had an equivalent method.

Operating without a firm foundation, the calculus was susceptible to challenge. The only test of validity for the new mathematics was to solve a problem for which everyone could agree on the answer, if not the solution. Even among the new generation, however, a consensus was not always easily reached or even achieved. For example, l'Hospital questioned Jakob Bernoulli's conclusion regarding the radius of curvature at the inflection points of higher-ordered parabolas, but he could not prove that his alternative was correct and therefore asked Huygens to settle the matter, presumably by resorting to his old-fashioned geometry. Likewise, when Huygens rejected Bernoulli's analysis of the sail curve, in fact, Huygens was wrong (it was a matter of slow convergence, in modern terminology), but no one could satisfactorily answer his objection, although l'Hospital made some attempt to do so.[20]

Huygens was right when he claimed that he could solve all problems with his own methods and that the calculus had many unresolved difficulties. Yet Leibniz was also right when he claimed that the algorithmic nature of the calculus gave it an efficacy that went beyond the immediate problem. Gradually the power of the new method became evident to Huygens. When, in the process of analyzing the curve, both Leibniz and Johann Bernoulli reduced the construction of the catenary to the quadrature of the hyperbola, Huygens realized that he had been left behind, for although he could confirm their result using his own method, the fact did not follow automatically from his derivation as it had from theirs.[21] In addition, although quite obviously the lesser mathematician, l'Hospital, applying the calculus rather than his imagination, was able to rectify the exponential curve that Huygens himself had introduced to the mathematical world.[22] Finally, at the very close of his life, Huygens ruefully set about to study all those explanations that Leibniz had been plying him with.

Like his mathematics, Huygens's physics was transitional, which means that here too he was eventually left behind. He moved beyond the rigid systems of the past, including Cartesianism, but he did not create a replacement; indeed, he seems to have been repulsed by the idea. He was content to solve problems, and for that task his geometric analysis was perfectly matched by his kinematics. This characterization of Huygens as a problem solver has come to have a negative connotation. He failed because he did not move beyond isolated problems to create a new physics, just as he failed to move beyond specific techniques to create a new mathematics. Yet this prejudice masks the very essence of his contribution. By focusing on the problem and demanding that any general system include its specific, technical solution, Huygens helped to move science toward the rigorous discipline that it is envisioned to be today. His problems came from nature; his solutions came from mathematics.

Perhaps the greatest tribute to Huygens's ability is the tendency of commentators to fault him for not having developed, or at least accepted, the calculus and modern Newtonian dynamics. What other scientist is blamed for not having discovered the inventions of a succeeding generation? Although only thirteen years separated him from Newton and seventeen from Leibniz, it was truly a generational gap, evident when one remembers that Huygens contributed heavily to the one textbook they did study, van Schooten's *Geometria*. Certainly, his contemporaries like Wallis (1616–1703), who provided the other influential textbook of the mid-seventeenth century (*Arithmetica Infinitorum*), and Wren (1632–1723) are not scolded for their failure to keep up.

Nevertheless, it is also true that Huygens was temperamentally unsuited to be a revolutionary. His life, intellectually and socially, was too amenable. Though he is often portrayed as being lonely and aloof, for most of his life he moved in very high circles: the Orange court at home, the intellectual elite at Montmor's, the Sun King's court as it moved into its prime. There is little evidence that he felt estranged from this society, and certainly he did what would please and win favor, working on topics of fashionable interest such as carriages and fountains for Versailles. His position at the Académie Royale and the Baconian ideals it drew out of him reinforced his conservative tendencies.

He was particularly reluctant to create a grand synthesis around which a school could gather. Picking up on Huygens's own remark concerning Boyle's failure to formalize his view of nature before dying, Leibniz continually pressed Huygens to compile his manuscripts and publish a unified treatise, but Huygens resisted the entreaties. His constant battles with stubborn advocates of one master or another, especially the Cartesians, had prejudiced him against system builders and their sycophants.[23] Even his most speculative work, the *Cosmotheoros,* ends with a slap at Descartes and a statement that mankind is far from knowing how things are and will never know how things formed and evolved.[24] The desire to be a new Descartes was Leibniz's greatest fault in his eyes and probably contributed to Huygens's distaste for the calculus.[25]

Of course, he could also find more trivial excuses for avoiding a new treatise. When Leibniz persists in pushing for the grand synthesis, Huygens reminds him that he has not even read and responded to the *Traité de la lumière,* so why write more. Moreover, l'Hospital is keeping him too busy with mathematics for him to do physics. In the meantime, to l'Hospital he explains that he must work on the physics that he has promised to publish.[26] His self-appointed disciples could only continue to cajole and to hope.

His personality, his concentration on problems over technique, his need to retain the physical tie, his distaste for heroes – all conspired to lessen his impact. Yet his influence on those who would lay the foundations of modern applied mathematics, particularly differential geometry and rational mechanics, cannot be denied. However much Huygens might rail against their worthless generalities and their unsubstantiated methods, Leibniz, Tschirnhaus, the Bernoullis, and l'Hospital were his heirs, a fact they would quite readily concede. Unfortunately, the true measure of Huygens's influence is difficult to assess because, aside from Leibniz, his followers have been as neglected by historians of science as has the master himself. Certainly, he did not dominate their work the way Newton did the British who followed him, but the sense of esteem that comes through in their letters has a basis that extends beyond politeness. If nothing else, the work Huygens did for the *Horologium Oscillatorium* showed that a complete and thorough mathematical analysis of a physical system was possible, and this encouraged them to expand mathematics into other areas of physics.

The symbiotic relationship between mathematics and physics that is the essence of applied mathematics is exemplified in Huygens's development of evolutes.[27] He did not impose mathematics on nature, as if it were some ideal form into which the untidy real world must be pressed. His mathematics grew along with his physics. Nor were Huygens's clocks models in the sense usually described by philosophers of science when they discuss analogy. That is, they were not static, completed entities whose familiar features could lead him to create his physics and/or mathematics by some kind of parallelism. Rather his horology, mathematics, and physics developed simultaneously and they cannot be pried apart. The pendulum prompted a mathematical question; the mathematical result provided a physical clue; the theoretical physics engendered a mechanical adjustment.

It was a beautiful, intricate display of creativity, all performed within the realm of, to use Leibniz's felicitous phrase, the Geometry of Archimedes. Indeed, Huygens spent his life confirming his right to the name with which Mersenne had christened him in his youth: the new Archimedes.[28] Yet at the time of his death, even that appellation seemed inadequate to Leibniz,[29] for whom he was simply:

the Incomparable Huygens.

Notes

In the following notes, references are cited by author and short title or, if there is no ambiguity, by the author's name alone. Full citations can be found in the Bibliography. References to the twenty-two volumes of the *Oeuvres complètes de Christiaan Huygens* are cited by *OH,* followed by the volume and page numbers; for example, Vol. 12, pp. 123–125 is listed as *OH* 12:123–5. References to the manuscripts, all of which belong to the collection *Codices Hugeniorum* housed at the University of Leiden, are cited by manuscript number and folio; for example, *Hug.* 26, f. 45r–47v. The editors of *OH* chose to refer to the manuscripts not by the library's numbers but by descriptive titles. In some cases this decision presents the modern scholar with no problems; such is the case with the workbooks, which Huygens himself lettered *A, B, C,* etc. In other cases this method of citation has created confusion; for example, *De Vi Centrifuga* is contained in the folder labeled *Chartae Mechanicae,* yet retains its own pagination in *OH.* In order to prevent confusion and to minimize my citations, I will refer to individual items by their place in *OH* if they were edited and by their proper manuscript number if they were not included. The accompanying table correlates all the pertinent information. It is arranged chronologically, according to my reconstruction of events; all dates are 1659 except where noted. The folio span under "*Hug.*" is sometimes larger than that listed under "MS Name" because the editors did not always transcribe the complete passage. *CMech* and *CMath* are abbreviations for *Chartae Mechanicae* and *Chartae Mathematicae,* respectively.

1. INTRODUCTION

1. In modern terminology, the evolute of a curve is the locus of its centers of curvature; Huygens's definition will be given later.
2. The best readily available biography of Huygens is Bos's entry in the *Dictionary of Scientific Biography.* In 1979 two international congresses celebrated the 350th anniversary of Huygens's birth, and the resulting proceedings, although inadequate as comprehensive biographies, are the most complete surveys to date of his work; see the Bos et al. and Centre National de la Recherche Scientifique editions. Most of the other biographies listed in the Bibliography are woefully inadequate and are listed only for completeness. Finally, there are the voluminous biographical notes written by Huygens's last editor, J. A. Vollgraff. A collection of his notes in French constitutes the biography of *OH* 22; an

Short description	Date	*Hug.*	MS Name	*OH*
Empirical plates	1657	5 f. 26r	Book *K* p. 45	17:17–20
Cycloid rectified	1/11	10 f. 34r–35r	Book *A* pp. 67–9	14:363–7
"Cartesian" passage		26 f. 78r	*CMech* f. 78	17:276–7
Mersenne exp., trial 1	10/21	10 f. 79v	Book *A* p. 158	17:278
De Vi, first draft[a]	10/21	26 f. 4r–9v	*De Vi* pp. 1–12	16:255–311
Mersenne exp., trial 2	10/23	10 f. 79v	Book *A* p. 158	17:278
Galilean treatise, beginning		10 f. 80r–v	Book *A* pp. 159–60	17:278–81
Galilean treatise		10 f. 80v–86v	Book *A* pp. 160–72	17:125–37
Relation of time to length		10 f. 87r	Book *A* p. 173	16:319
Preliminary conical clock		10 f. 87r		Not edited
Conical clock[b]	10/5	10 f. 87v–88r	Book *A* p. 175	17:88–91
De Vi, revision[c]		26 f. 10r–17v	*De Vi* pp. 13–28	16:256–89
Mersenne exp., trial 3	11/15	10 f. 88v	Book *A* p. 176	17:281
Calculation of $r = 9\%_{20}$		10 f. 130v		Not edited
Fall, infinitesimal arc[d]	12/1	26 f. 72r	*CMech* 72r	16:392–7
Lemmas for 72r		26 f. 73r	*CMech* 73r	16:398–400
Fall, quarter arc		26 f. 73v	*CMech* 73v	18:374
Last lemma		26 f. 74r	*CMech* 74r	16:401–3
Cycloidal to conical		10 f. 89r	Book *A* p. 177	16:320–2
Calculation of $r = 9\frac{1}{2}$		26 f. 72v		Not edited
Compound pendulum		10 f. 89r–v	Book *A* pp. 177–8	16:385–91
Evolute of cycloid		25 f. 158v	*CMath* 158–9	17:145
Improved proof		25 f. 158r, 159v	*CMath* 158–9	17:143–4
Involute of cycloid		26 f. 75v	"Loose sheet"	14:404
Evolute of parabola		26 f. 75r		Not edited
Evolute of ellipse		10 f. 91r–93v	Book *A* pp. 181–5	14:387–90
Isochronism, proof 2		10 f. 94r	Book *A* p. 187	16:404–5
Isochronism, proof 3	12/15	10 f. 94v–96r	Book *A* pp. 188–91	16:405–12
Evolute of hyperbola		10 f. 96v–101v	Book *A* pp. 192–202	14:391–8
Results axiomatized		10 f. 102r–104v	Book *A* pp. 203–8	14:399–403
Preparing new edition	1660	10 f. 106r	Book *A* p. 211	17:100
Astronomy		10 f. 106v–113r	Book *A* pp. 212–25	15:535–50
Improved proof		10 f. 113v	Book *A* p. 226	14:405
Outline		10 f. 113v	Book *A* p. 226	17:120–3
Higher-order conics		10 f. 117v–118r	Book *A* pp. 231–2	17:147
		10 f. 122v–123r	Book *A* pp. 244–5	17:146
Making a cycloid		10 f. 125v–127r	Book *A* pp. 250–2	17:98–100

[a] Originally in *Hug.* 10 between ff. 79 and 80; portion of text still visible.
[b] See notes on date.
[c] Originally in *Hug.* 10 between ff. 88 and 89; contains figure for work on f. 89r.
[d] Foliated backward; originally in *Hug.* 10 before f. 89, on which calculations from this sheet continue.

equally massive set of his notes in English exists in a manuscript now on deposit with the University of Leiden library (MS BPL 2781). Elzinga attempted to organize this manuscript into a biography (*Notes*), adding his own philosophical interpretation of Huygens's role in seventeenth-century science (cf. his *On a*

Research Program). I am grateful to the author for providing me with copies of both his works.

3. In fact, his work began in December 1656, but a model was not constructed until 1657. Huygens's early manuscripts on horology are collected in *OH* 17.

4. The question of priority is complicated by the conflation of the pendulum per se with the pendulum-regulated clock. Many before Huygens, including Galileo, Mersenne, and Riccioli (see Chapter 2), used the regularity of a swinging pendulum as a timing mechanism. The crucial problem, as Defossez (who sides with the partisans of Huygens) sees it, is to keep an accurate record of the time that the pendulum beats, a task that requires mounting the pendulum on a clock. Deflating the claims for Leonardo's priority, Bert S. Hall has argued similarly that turning the theoretical pendulum into an accurate physical mechanism involves overcoming substantial conceptual as well as technical hurdles; see his review of the da Vinci *Madrid Codices*.

5. The *Horologium* (The Hague, 1658) is reprinted with facing-page French translation in *OH* 17:41–73. Huygens's clocks exhibited many mechanical variations (some introduced by his clockmakers) that I do not intend to describe except where they manifest his theory. Thus, for example, I will not be concerned with the placement of the verge, or even the type of escapement mechanism. However, since it is not obvious from his diagrams, I should mention that his pendulum clocks were small mural or mantle clocks and not large, floor-standing grandfather clocks, which were an English development. Also, although I will always describe his pendulum as if it consisted of a spherical bob mounted on a ropelike cord (a physicalization of the ideal simple pendulum of point-mass bob and dimensionless cord), in fact, his pendulum usually consisted of a short, fine chain or other flexible medium joined to a rod that then led to the bob, which could be one of many shapes.

6. The *Horologium Oscillatorium sive de motu pendulorum ad horologia aptato demonstrationes geometricae* (Paris, 1673) is reprinted with facing-page French translation in *OH* 18:70–368. A facsimile reprint is available (London, 1966). In addition to Blackwell's English translation, a German translation was published in *Ostwald's Klassiker der Exakten Wissenschaften,* No. 192 (Leipzig, 1913).

7. I use the term "isochronous" in the loose way that most people, including Huygens, apply it. A clock is isochronous if its pendulum beats isochronously; that is, it maintains a uniform rate of swing irrespective of the amount of deflection, which theoretically is possible only if its bob traces out an isochronous curve. This curve is sometimes called the tautochrone, the isochrone being defined by Leibniz as the curve of uniform vertical descent (the semicubical parabola); compare Archibald's definitions with James and Beckenbach's. Adding to the confusion, modern European commentators use "tautochronous."

8. Except for Volumes 1 to 10, which contain Huygens's correspondence, the collected works are arranged topically, not chronologically, and consequently the interaction between Huygens's mathematics, physics, and horology is not readily apparent. On the whole, I have found the editorial *avertissements* that precede most major sections of his manuscripts to be inadequate and in a few

cases misleading. Since I view the commentaries as overviews and not definitive interpretations, I do not intend to note differences between their explanations and mine, except in cases where I feel the importance of the material warrants it.

9. James Gregory also had a theory of "evolution" and "involution" in his *Geometriae Pars Universalis* (Padua, 1668), but it has no relationship to Huygens's theory. Gregory "unrolled" or fanned out a curve by a transformation akin to mapping with polar coordinates. For a description of his theory, see Baron, pp. 231–4.

10. As is the case with most ideas, the concept preceded the terminology in Huygens's manuscripts. In order to emphasize that the birth of the concept is being described and recognizing that people tend to transpose the terms "evolute" and "involute," I will avoid using the technical terminology until the idea is well developed (Chapter 5). The French equivalents of the two terms are *développée* and *développante*.

11. Huygens proves that only one involute can cross the evolute at any given point. Thus, the involute described by evolution that goes through the given point must be the same as the involute whose normals define the evolute mathematically (*OH* 18:188–99).

2. ACCELERATED MOTION: GRAVITY

1. The list of recommended readings compiled by Huygens's first teacher, Stampioen de Jonge, includes the works of Stevin, Kepler, Vitello, Apollonius, Descartes, Tycho, Copernicus, Clavius, and Viète (*OH* 1:5–10).

2. Albert Girard, *Les oeuvres mathematiques de Simon Stevin de Bruges* (Leiden, 1634), p. 508. As his editors note, Huygens refers to Girard's edition in a 1691 letter to Leibniz concerning the catenary, citing Girard's mistaken assumption and claim of proof and also citing his own refutation, which he states he sent to Mersenne and also showed Descartes (*OH* 10:187–8).

3. *OH* 11:37–44. Huygens's whole discussion is very much in the tradition of Stevin and as such is further proof of the connection of his work with a reading of Girard's edition of Stevin.

4. *Sublimium ingeniorum crux. Jam tandem aliquando deposita a Joanne Caramuel Lobkowits, Gravium lapsum cum tempore elapso componente, concordiamque experimentis & demonstrationibus Geometricis firmante* (Louvain, 1644).

5. *OH* 11:68–72. Huygens's various proofs of the times-squared law are compared by Yokoyama.

6. "Outre cecy j'ay demonstré que s'il est jetté de quelque costé, qu'il descrit une parabole; de tout cecij et encor d'une infinité de choses qui en dependent je n'aij jamais sçeu la demonstration avant que de l'inventer moij mesme" (*OH* 1:18–19).

7. "Probare institueram, projecta pondera sursum vel in latus, parabolam describere, sed interea temporis in manus incidit libellus Galilei de motu accelerato naturaliter et violento; quem cum videam haec et plura alia jam demonstrasse, nolui Iliada post Homerum scribere" (*OH* 11:72–3). Given his father's

supposedly great admiration for and knowledge of Galileo's works, it is surprising that Christiaan had not read the *Discorsi* at an earlier date.

8. Constantijn forwarded to Mersenne a copy of Christiaan's letter to his brother (*OH* 2:549, Sept. 12, 1646).

9. "Comme i'honore grandement Monsieur vostre pere, et que ie croy luy faire plaisir de vous parler de vos propositions dont vous dites auoir la demonstration, ie vous diray seulement sur la derniere, que ie ne croie point que vous en ayez la demonstration, si ie ne la voy; dont voicy ma raison. Les graues qui tombent ne vont pas tousiours augmentant leur vitesse suiuant les nombres jmpairs. . . .

"J'aioute que les principes que Galilee a pris dans tout ce qu'il a dit du mouuement, ne sont guere fermes" (*OH* 1:558–9, Oct. 13, 1646). However, Mersenne did accept the times-squared law as being experimentally verified by his own work.

10. "I'attens auec grand desir la demonstration de Mons. vostre fils sur la proportion des cheutes des corps pesans, car il l'aura peut estre prise d'un biais jndependent des supositions de Galilée" (*OH* 1:22, Oct. 12, 1646).

11. Requiring the distances to be in constant proportion irrespective of the unit of time chosen means finding distances such that $(w+x)/(y+z)$, where w, x, y, and z are distances successively traversed in equal time, beginning from rest. Huygens found that

$$\frac{1}{3} = \frac{1+3}{5+7} = \frac{1+3+5}{7+9+11} = \cdots$$

is the only solution. See *OH* 1:24–8 (Oct. 28, 1646) for his letter to Mersenne. This text, written in French, is very similar to the original Latin derivation (*OH* 11:68–72). Without necessarily citing Huygens, commentators who deny that Galileo achieved the times-squared law experimentally have argued that he could have derived it in a way similar to this.

12. "I will end here for fear of detaining you any longer, and will send you in another letter my demonstration that a hanging cord or chain does not form a parabola, and what the tension on an ideal or gravity-free cord ought to be in order to form one; for which I also found a demonstration not too long ago" ("Ie fineray icij de peur de ne vous detenir pas trop longtemps, et vous envoyeray par une autre lettre la demonstration de ce qu'une corde ou chaine pendue ne faict point une parabole, et quelle doit estre la pression sur une corde matematique ou sans gravité pour en faire une; d'ont j'ay aussi trouve la demonstration, il n'y a pas longtemps"; *OH* 1:28, for the letter to Mersenne; 34–44, for a reconstruction of his proof). It has become traditional to claim that Mersenne set the problem of the catenary for Huygens after having read Galileo's assertion that it is a parabola; for example, see Bell's biography, p. 22.

13. "Ie vous asseure que i'ay si fort admiré la gentillesse de vostre demonstration des cheutes, que ie croy que Galilée eust esté raui de vous auoir pour garand de son opinion" (*OH* 1:30–1, Nov. 16, 1646). However, he does go on to say that he still has a few reservations about Huygens's explanations.

14. "I cannot agree to that, and concur with the Philosophy of Descartes, who among other laws of nature has declared that everything continues its motion

with the same speed that it had at the beginning, if nothing else impedes it" ("Ie ne puis pas consentir a cela, et m'en rapporte a la Philosophie de Monsr. des Cartes, qui entre autres loix de la nature a remarqué cellecy, à scavoir que toute chose continue son mouvement de las mesme vistesse que luy a esté donée unefois, si quelque autre chose ne l'empesche"; *OH* 1:24).

15. "Nemo autem satis abstracté motum consideravit, dum considerat motum lapidis aut sphaerae metallicae per aerem ex alto cadentis; maximè enim cum experientia conveniret sententia Galilaei nisi aeris resistentia id impediret. Sic ergo motum acceleratum melius considerabimus" (*OH* 11:69).

16. *De Motu Corporum ex Percussione* was published posthumously and does not have a formal introduction. However, some of the preliminary pieces contain general comments that appear to be notes for an introduction, including extensive quotations from Galileo (*OH* 16:112-13). The discussion quoted by Huygens occurs in the Fourth Day of the *Discorsi;* see the Drake translation, p. 242. In fact, Galileo did write a dialogue on percussion, but it was not published in the seventeenth century; it does appear as an added day in Drake's translation, pp. 281-303.

17. For Mersenne's letter, see *OH* 1:559. For Huygens's work on the cycloid in response to Pascal's challenge, see *OH* 14:347-78.

18. For Mersenne's attempts to interest him, see *OH* 1:22-3, 45-7, 50-6. For Huygens's eventual results, see *OH* 18:243-359.

19. For Mersenne's description of his early work, see *Harmonie universelle* (Paris, 1636), pp. 136-8, 220; *Cogitata physico-mathematica, Phenomena ballistica* (Paris, 1644), pp. 38-40; *Harmonicorum Libri XII* (Paris, 1648), *liber secundus, de causis sonorum,* pp. 20-4.

20. Chapter 19, pp. 152-9, of *Reflexiones Physico-mathematicae,* printed in *Novarum Observationum Physico-mathematicarum,* Vol. 3 (Paris, 1647).

21. The quarter arc of a circle of 3-foot radius is $3(\pi/2)$ feet, which is approximately 5 feet. Galileo's times-squared law for free fall can then be applied to compare the times of free fall through this distance and the radius, yielding

$$\frac{\text{time of free fall through arc length}}{\text{time of free fall through radius}} = \sqrt{\frac{3(\pi/2)}{3}} = \sqrt{\frac{\pi}{2}}.$$

Mersenne vacillates on whether the ratio of times is equal to or greater than the root, but at least he has a lower bound on the ratio of times.

22. Thus, Mersenne has lower and upper bounds for the ratio of times, namely $\sqrt{\pi/2}$ and $\sqrt{2}$, since

$$\sqrt{\frac{\pi}{2}} \leq \frac{\text{time on quarter arc}}{\text{time through radius}} < \frac{\text{time along chord}}{\text{time through radius}} = \sqrt{2}.$$

Huygens would show that, for small amplitude of swing and for a cycloidal arc of any amplitude, the ratio of times is $\pi/2\sqrt{2}$. Huygens's value actually violates Mersenne's lower bound, since he restricts the swing to small arc versus the quarter arc, thereby subtly but significantly changing the problem.

23. Under Mersenne's first assumption, the length of the quarter arc of the circle should equal, or at least not exceed, the free-fall distance, in this case 3 feet. Thus, the pendulum's length, which is equal to the radius of the circle, should be $3(2/\pi)$ feet, which is approximately 22 inches.

24. The problem inherent in Mersenne's experiment, as Huygens was to discover, was the difficulty of determining the simultaneity of the sounds of impact.
25. Giambattista Riccioli's experiments on gravitational fall, including comparisons of the fall of heavy and light objects through air and water and down a 35-foot inclined plane, are discussed in his *Almagestum Novum* (Bologna, 1651), Book 9, Section 4, Chapter 16, pp. 381–97. For his discussion of the odd-numbers law and his description of the pertinent experiments, performed in the early 1640s, and the resulting table, see pp. 386–7.
26. Only the second trial actually has a direct measure of the fall in 1 second, 15 feet. For the first trial, a computation yields 14.4 feet; for the third, 15.33 feet.
27. Passage quoted as printed; p. 234.

3. ACCELERATED MOTION: CENTRIFUGAL FORCE

1. "Expertus 21 Oct. 1659. Semisecundo minuto cadit plumbum ex altitudine 3 pedum et dimidij vel 7 pollicum circiter. Ergo unius secundi spatio ex 14 pedum altitudine" (*OH* 17:278). The pendulum's length of 3 feet 1 inch is noted in the drawing that accompanies this description. In the *Horologium Oscillatorium* Huygens does note the difference in units, stating that 139 Parisian Royal feet are equal to 144 Rhenish feet (*OH* 18:351). Both lengths are longer than the current U.S. foot.
2. *OH* 16:255–311. The manuscript consists of a series of sheets numbered 1 through 28, the first page being dated Oct. 21, 1659. Instead of reproducing the manuscript as it exists, the editors of Huygens's collected works have followed the posthumous edition of *De Vi Centrifuga* published in 1703 by Volder and Fullenius, who used the list of propositions on circular motion in Part 5 of the *Horologium Oscillatorium* as the basis for their ordering of the contents. The beginning pages of the manuscript not used by Volder and Fullenius have been reproduced in *OH* as appendixes to *De Vi Centrifuga*. Internal evidence reveals that the manuscript was not completed on Oct. 21, but only begun then, for the later section (that used in the posthumous version) is obviously a redaction of the earlier part, and different constants of gravitational acceleration divide the work further, as I will discuss.

    Thus, there are three versions of *De Vi Centrifuga*: (1) the original list of propositions, numbered 1 through 20, occupying MS pp. 1–12; (2) the revision made by Huygens, MS pp. 13–28; (3) the edition of Volder and Fullenius, which uses Huygens's revision plus propositions 18 to 20 from the first draft and adds propositions on the conical and paraboloidal pendulums (with proofs supplied by the editors) listed in the *Horologium Oscillatorium*, Part 5, but not developed by Huygens at this time. The manuscript is now contained in the collection titled *Chartae Mechanicae* (*Hug.* 26).
3. "Descensum gravium per 1, 3, 5, 7, &c. comprobat Ricciolus lib. 9. De mensura certa per horologij oscillat" (*OH* 16:303). "Grave item conatus progredi" also appears on the manuscript but not in *OH*.
4. "Hoc idem est gravitas corporis quod conatus materiae ipsi aequalis ac celerrimè motae a centro recedendi. Qui sustinet suspensum is detinet materiam istam ne recedat, qui vero sinit cadere, eo ipso facultatem praebet eidem materiae recedendi a centro secundum radium, cum autem a principio recedat à

centro secundum numeros impares ab unitate, id facere non potest quin corpus grave cogat similiter accelerato motu versus centrum accedere. adeo ut haec initio motus necessario aequalia sint, recessio materiae a centro, et corporis cadenti accessus versus centrum. Unde et comperto descensu hujus qui certo tempore contingit, velut si 1‴ decidit per spatium ⅗ lineae, cognoscemus quoque ascensum materiae istius a centro, qui nempe 1‴ tempore etiam erit ⅗ lineae.

"Hinc jam celeritas materiae data terrae semidiametro.

"Hinc porro vis centrifuga in minoribus circulis. Videndum enim qua in re consistat et quid determinet magnitudinem istius conatus. Nempe quantitas recessus certo tempore. haec vero et a celeritate gyrationis et a magnitudine circuitus pendet" (*OH* 17:276-7).

As the editors explain (p. 277, n. 6), three-fifths of a line in ¹⁄₆₀ second corresponds to a fall of 15 feet per second, Riccioli's value. Huygens actually makes the necessary calculation on the manuscript:

$$(15)(12)(12)/(60)^2 = \tfrac{6}{10} = \tfrac{3}{5}.$$

5. For examples of this interpretation, see Elzinga, *Notes,* p. 136; Westfall, *Force,* pp. 172-3; as well as the editorial comments in *OH* 17:244.
6. More than thirty years later, the same explanation of weight would appear in Huygens's *Discours de la cause de la pesanteur,* with the direction of causality at least superficially solved: "The effort to elongate from the center is thus a constant effect of circular motion. And although this effect appears directly opposed to that of gravity...this same effort, which the circulating bodies make to move away from the center, is the cause by which other bodies move toward the same center.... Thus, it is in this that the weight of bodies truly consists: of which one can say, that it is the effort that the fluid matter, which rotates circularly around the center of the earth in all directions, makes to elongate from the center, and to push in its place the bodies that do not participate in this motion" ("L'effort à s'eloigner du centre est donc un effet constant du mouvement circulaire. & quoyque cet effet semble directement opposé à celuy de la gravité...ce mesme effort, que font les corps tournants en rond à s'eloigner du centre, est cause que d'autres corps concourrent vers le mesme centre.... C'est donc en cela que consiste vraisemblablement la pesanteur des corps: laquelle on peut dire, que c'est l'effort que fait la matiere fluide, qui tourne circulairement autour du centre de la Terre en tous sens, à s'eloigner de ce centre, & à pousser en sa place les corps qui ne suivent pas ce mouvement"; *OH* 21:452, 456).
7. Huygens's editors attribute his approach to a reading of Descartes (*OH* 17:244). Certainly, Huygens's workbook gives no clue as to what is to follow. The topics of the pages that precede his Oct. 21 experiment include geometric properties of triangles (*OH* 16:379), the tonal quality of a bell (*OH* 17:339), and drawings for a lantern slide show in which a cadaver takes off its head and tosses it in the air (*OH* 22:197).
8. Though trying to balance its similarities with the work that follows and the noticeable differences is a challenging mental exercise, the dating of this manuscript is not crucial. Indeed, my argument, which emphasizes the differences

and the consequent importance of Mersenne, is stronger if Huygens actually wrote the passage earlier and therefore abandoned the problem of gravity until Mersenne's experiment renewed his interest.

9. "Porro quoties duo corpora aequalis ponderis unum quodque filo retinetur, si conatum habeant eodem motu accelerato, & quo spatia aequalia eodem tempore peractura sint, secundum extensionem fili recedendi: Aequalem quoque attractionem istorum filorum sentiri ponimus, sive deorsum sive sursum sive quamcunque in partem trahantur. . . . Idque in initio motus tantum spectandum est, accepta parte temporis quamlibet exigua. . . . Nunc videamus quis quantusque conatus sit corporibus filo vel rotae quae circumgyratur alligatis, ut à centro recedant" (*OH* 16:259).

10. Huygens's changing values for the constant of gravitational acceleration help one to arrange the works chronologically. The "Cartesian" passage uses 15 feet, thus predating Huygens's repetition of Mersenne's experiment; the first draft of *De Vi Centrifuga* uses 14 feet in its fourth proposition, consequently postdating the experiment that yielded this value.

11. *OH* 16:297–9.

12. For a parabola with a *latus rectum* (the *constant* that when multiplied by the *y* value gives the square of the *x* value) equal to *BG*, the defining equation is $CB^2 = CF \cdot BG$ (see Fig. 3.2). In the circle, *FA* is the mean proportional of *BA* ($= CF$) and *AG*, and thus $FA^2 = BA \cdot AG$ ($= CF \cdot AG$). If *A* is close to *B*, then $BG = AG$ and $FA = CB$, making the two equations interchangeable. Actually, Huygens's approximation in *De Circuli Magnitudine Inventa* was more general in that he was concerned with fitting a parabola to a circle at a point other than the vertex (*OH* 12:165).

13. The diagram and formula are reproduced in *OH* 16:302. My reconstruction is based on the proofs of *OH* 16:261–5. Also see *OH* 16:297–9, where Huygens uses the construction under discussion to prove his variant of Mersenne's problem, namely that the tension on the cord of a circular pendulum at the bottom of its fall through the quarter arc of the circle is three times the tension felt if the body were simply suspended.

14. Huygens provides another method for deriving the same formula in the revised *De Vi Centrifuga* when he shows that the uniform circular speed necessary for a body's centrifugal force to equal its weight is the terminal velocity of fall through one-fourth the diameter of the circle (*OH* 16:275–7).

15. *OH* 16:303. The editors claim (pp. 303–4, n. 8) that Huygens knew the absolute magnitude of centrifugal force, equivalent to the formula $F = mv^2/r$. Yet it is only in the propositions that follow that Huygens develops the proportions equivalent to the formula. They also erroneously claim that these first four propositions are more general than the case in which centrifugal force equals weight. It is precisely because he is dealing with this case of equilibrium that he can find the radius of a seconds-circle from his known constant of fall.

16. *OH* 16:304. The distance traversed in free fall is proportional to the square of the time, so doubling the time would quadruple the distance $\pi^2 d$ and thus *d* is quadrupled.

17. *OH* 16:304. Setting $14 = \pi^2 d$, where $\pi = 22/7$, gives *d* approximately equal to 17 inches. Thus, the radius would be 8.5 inches. In 24 hours a body would fall

14(3,600 · 24)² feet, since distance is proportional to the square of the time, and therefore for the centrifugal force to equal gravity the radius of the earth would have to be ½ · (⅟₂₂)² · 14 · (3,600 · 24)², or 5,290,250,579 Rhenish feet, compared with Snell's value of 19,595,160 feet. At first Huygens records only the two values, but when he returns to enter a new constant of acceleration, he recomputes his radius (5,164,292,231), divides by Snell's value, and thus finds that they differ by a factor of 265.

18. *OH* 17:278. The new length for the pendulum follows from the times-squared law, for it should equal (1½)² times 3 feet 1 inch. Actually, Huygens determined that a ball would fall 13 feet 7½ inches in 1 second, but rounded this value off to 13 feet 8 inches.

19. *OH* 16:306.

20. *OH* 16:306. In other words, the force *D* necessary to prevent the body from sliding down the plane is equal to the weight *C* times the tangent of angle *BAF*, because the sustaining force must equal in magnitude the horizontal component of the normal force, so that *C* and *D* are in equilibrium with the normal force. This principle and the pulley technique used throughout this work are found in writings attributed to the school of Jordanus Nemorarius and were subsequently refined by Stevin, who was undoubtedly Huygens's source.

21. *OH* 16:306–7. From the value obtained on Oct. 23, a body falls 13 feet 8 inches in 1 second; thus, the diameter equals 13 feet 8 inches times (⅟₂₂)², which is approximately 16⁶⁄₁₀ inches, making the radius 8³⁄₁₀ inches. Later Huygens alters the value to the 9½ inches that is shown on the figure.

22. *OH* 16:307. In the parabola, $x^2 = 2py$, the tangent at any point equals x/p, where $2p$ is the *latus rectum* and in this case $p = DA$. At $y = p/2$ (i.e., one-fourth the *latus rectum*), x equals p (or $-p$) and the tangent is 1, yielding an angle of 45°. In general, the tangent at *H*, *HG/GF*, equals *HK/DA*.

23. *OH* 16:307–8.

24. "Minimo motu calicis, ita ut vertex ejus circellum exiguum describat, continuari potest motus globuli; cujus si circuitus numerentur, exacta temporis mensura hoc pacto habebitur, pendulo accuratior" (*OH* 16:308).

25. *OH* 16:309–10. In the process of deriving the present formula Huygens has followed the steps necessary to achieve the equation $t = 2\pi\sqrt{h/g}$, where h is the height of the cone. He begins with the equivalent of the formula $t = 2\pi\sqrt{r/g}$, where r is the radius of a circle on which centrifugal force equals the weight. In this case, $r = GA$. He now shows that, in general,

$$\frac{\text{time of revolution on a cone of height } HA}{\text{time of revolution on a cone of height } GA} = \frac{\sqrt{HA}}{\sqrt{GA}},$$

so that the expression for the time of revolution on a cone with altitude *HA* is $t = \sqrt{HA/GA} \cdot 2\pi\sqrt{GA/g} = 2\pi\sqrt{HA/g}$. Since Huygens almost always writes formulas in proportional terms, he does not expressly give this equation and g does not directly appear in his relations.

26. *OH* 16:310–11. Compare with Stevin's third corollary to Proposition 19, Book 1, *De Beghinselen der Weegconst* (Leiden, 1586); reprinted in his *Works* 1:180–1. In a modern derivation, using the parallelogram law of forces, the tension is $\sqrt{(\text{wt. } H)^2 + (BF/FE)(\text{wt. } H)^2}$, or $(\text{wt. } H)\sqrt{1 + (BF^2/FE^2)}$, which is

(wt. *H*)(*BE/FE*). But *BE/FE* is equal to *AB/AD* by similar triangles. Therefore, the tension is equal to the secant of the angle subtended times the weight of the bob.

27. The drawing appears in *Hug.* 10, f. 87r; not reproduced in *OH*.

28. *OH* 17:88–91. The chain is replaced in another drawing by a hydraulic piston.

The date on the manuscript is Oct. 5, 1659. However, there are good reasons for believing that Huygens intended Nov. 5. The page (f. 88r) in his workbook (*Hug.* 10) on which the drawing appears is after that on which he records his experiments of Oct. 21 and 23 (f. 79v) but before that of Nov. 15 (f. 88v). In the text under question Huygens states that a conical pendulum of about 6 inches will make 5,040 revolutions per hour when kept at an angle of 45°, and in his calculation of this value, which appears on f. 87v and is not reproduced in *OH*, he clearly uses the Oct. 23 value of the constant of gravitational acceleration, namely 13 feet 8 inches. In addition, if Huygens had already constructed the conical pendulum by Oct. 5, it is strange that he would not use it in his attempts to find the constant of gravitational acceleration on Oct. 21 and 23, for he does refer to its use on Nov. 15. Of course, the date could refer to his first conception of the conical pendulum, of which this manuscript is a revision with added calculations.

Huygens's editors do not question the date. They are then pressed to explain the value of 6 inches, when on Oct. 5, Huygens's value for the constant of gravitational acceleration would have been Riccioli's 15 feet and the length therefore should be closer to 6.6 inches. They conclude that the 6 inches was merely a "gross observation." By accepting the early date, they must also conclude that Huygens had most of his theory of centrifugal force before Oct. 5 and thus before any of the extant manuscripts; see *OH* 17:91, n. 4, and their long revision of this note, *OH* 19:682–3, as well as *OH* 22:514.

The interpretation of the text is also a difficult matter, since Huygens's wording is so unclear. "When the length *BA* is shortened, as much weight ought to be added to both the globe *A* and to the weight *E* (taken by itself) as would hang a part of the chain equal to the length of that part by which the length of the cord was shortened" ("Ideo cum minuitur longitudo *BA*, tantum gravitatis et globo *A* et ponderi *E* singulis est adjiciendum quantum pendet particula catenae aequalis longitudine ei particulae qua fili longitudo diminuta est"). Yet clearly he cannot keep adjusting the weight of *A* when it is in motion, and the rising and falling of the chain are meant to keep equilibrium – hence, my interpretation.

29. The gears determine the number of rotations per hour, because with each rotation the escapement allows one tooth of the crown wheel to advance, and therefore the combination of gears will return to its starting position in 1 hour only if the escapement's release is properly timed through an adjustment of the pendulum's length. The values 5,040 ($= 12 \cdot 15 \cdot 14 \cdot 2$) and 4,320 ($= 12 \cdot 15 \cdot 12 \cdot 2$) are obtained by multiplying the number of teeth in each successive gear, which gives the number of rotations the pendulum must make in order that the system return to the starting position. Huygens computed the number of revolutions for other combinations of gears but did not proceed to calculate the corresponding lengths for the pendulum, probably because the

combinations did not approximate a rotational time of ¾ second (4,800 revolutions) as closely.

30. The calculations appear in *Hug.* 10, f. 87v; not reproduced in *OH*. A fall of 13 feet 8 inches corresponds to a length of 11⁸⁄₁₁ inches for a 45° conical pendulum that rotates in 1 second (3,600 rotations per hour). Thus, for 4,320 rotations per hour the length is $(3,600/4,320)^2 \cdot (11⁸⁄₁₁)$, or 8½ inches.

31. Huygens's editors claim that it "appears possible" that the clock was never built but was only a "model" (*OH* 17:97). Their conclusion is in part a consequence of their acceptance of Oct. 5 as the date of the design. Editor Crommelin repeats the claim in "Clocks of Huygens," saying that it was "invented probably in 1659 or 1660 and constructed about 1667 or 1668" (p. 65).

Yet clearly some device was constructed and used to achieve a result *ex motu conico penduli*. Accepting the interpretation of the editors, Costabel has argued that, therefore, the device used by Huygens was the ball-in-rotating-tube apparatus described in the opening pages of the first draft of *De Vi Centrifuga* ("Isochronisme"). However, a few paragraphs after describing that device, Huygens himself rejects it as too unstable, depicting in its place the ball revolving inside a chalice. Considering the detailed calculations of gears and lengths that Huygens made for his design of the conical-pendulum clock, in contrast to the one-line descriptions of the devices in *De Vi Centrifuga,* I would rather take Huygens's words literally. Of course, the conical pendulum (with adjusting chain) need not have been mounted on a clock. However, Huygens would have had to count the rotations of the pendulum by some means, and I doubt that he would have performed the arduous task of counting revolutions over a prolonged period of time, even if aided by his brothers, in the manner adopted by Riccioli and his spiritual brothers. Huygens could have counted the number of revolutions in a short unit of time, such as 1 hour, where the hour was determined by a detached clock (already constructed and regulated by a simple pendulum), but his extensive calculations of gear combinations lead me to believe that he did mount the conical pendulum onto a clock especially constructed, or at least modified, for the occasion. The theoretical implications were too important for the experiment to be treated haphazardly.

32. That is, a pendulum rotating 4,320 times per hour should actually have a length of 9⅓ inches, not 8½ inches; 5,040 should correspond to 6.8 inches, not 6 inches. Or, adjusting rotations instead, an 8½-inch pendulum should make 4,620 rotations per hour; a 6-inch one, about 5,380.

33. "15 Nov. 1659. Pendulum *AB* semivibrationi impendebat ¾ unius secundi; filum idem *BDC* plumbum *B* et glandem *C* retinebat, deinde forficibus filum incidebatur, unde necessario eodem temporis articulo globulus *C* et pendulum moveri incipiebant. plumbum *B* in *F* palimpsesto impingebatur, ut clarum sonitum excitaret. globulus in fundum capsae *GH* decidebat. simul autem sonabant, cum *CE* altitudo erat 8 pedum et 7 unciarum circiter. Sed etsi 3 quatuorve uncijs augeretur vel diminueretur altitudo *CE* nihilo minus simul sonare videbantur. adeo ut exacta mensura hoc pacto obtineri necqueat. At ex motu conico penduli debebant esse ipsi 8 pedes et 9½ unciae. unde uno secundo debebunt peragi a plumbo cadente pedes 15⁶⁄₁₀ proxime.

Sufficit quod experientia huic mensurae non repugnet, sed quatenus potest eam comprobet. Si plumbum *B* et globulum *C* inter digitos simul contineas ijsque apertis simul dimittere coneris, nequaquam hoc assequeris, ideoque tali experimento ne credas. Mihi semper hac ratione minus inveniebatur spatium *CE*, adeo ut totius interdum pedis differentia esset. At cum filum secatur nullus potest error esse, dummodo forfices ante sectionem immotae teneantur. Penduli *AB* oscillationes ante exploraveram quanti temporis essent ope horologij nostri. Experimentum crebro repetebam. Ricciolus Almag. 1. 9 secundo scrupulo 15 pedes transire gravia statuit ex suis experimentis. Romanos nimirum antiquos quos a Rhenolandicis non differre Snellius probat" (*Hug.* 10, f. 88v). Huygens later replaced the values 8 feet 7 inches and 15⁶/₁₀ feet by 8 feet 9½ inches and 15 feet 7½ inches (= 15.625 feet), and it is the latter values that his editors reproduce in their transcription of the text (*OH* 17:281). They treat the first value of the constant of gravitational acceleration as a mere round-off of the second (p. 281, n. 6), whereas I find much more significance in the change.

34. A value of 8 feet 9.5 inches, or 105.5 inches, in ¾ second corresponds to 187⁵/₉ inches (= 187.555...), or 15.629 feet, in 1 second. Huygens probably calculated the value only to the tenths position. I have not found his derivations in the manuscripts that I have inspected.

    On *Hug.* 26, f. 72v (72r bears the date Dec. 1) Huygens begins a calculation to determine the distance fallen in the first second given a fall of 103 inches in ¾ second (i.e., the 8 feet 7 inches of his repetition on Nov. 15 of Mersenne's experiment). The calculation is not complete but is carried out far enough (18...inches) so that he could have realized the value would still be above 15 feet. Perhaps some doubt as to the exactness of his value still lingered and motivated his continuing search.

35. Newton also measured the constant of gravitational acceleration using a conical pendulum (probably not mounted on a clock); see Westfall, "Newton and the Acceleration of Gravity."

36. The proposition appears on p. 24 of the manuscript of *De Vi Centrifuga; Hug.* 26, f. 15v. Three successive values of the constant of gravitational acceleration appear: 13 feet 8 inches, 15⁶/₁₀ feet, and 15 feet 7½ inches. His editors downgrade the middle value, treating it as an insignificant round-off, just as they did with its appearance in the report of Nov. 15 (see note 33). Yet the number is accompanied by its own value for the radius of the seconds-circle (0.78 ft), and there is no reason to argue that Huygens did not accept it as a legitimate result.

37. Reprinted in *OH* 16:281–5; see note 2 regarding the various versions of the manuscript.

38. For Descartes's discussion of the circular motion of a sling see articles 57 to 59 of his *Principia Philosophiae* (Amsterdam, 1644), Part 3; reprinted in *Oeuvres de Descartes* 8:108–12.

39. See articles 60 to 63 in *Oeuvres de Descartes* 8:112–17. Aiton describes the Cartesian vortices in his third chapter.

40. See notes 5 and 6 of this chapter.

41. All citations refer to the Drake translation, *Dialogue;* see pp. 194–5 for the quotation, pp. 194–202 for the general discussion, although the entire Second Day is devoted to an examination of the effects of a spinning earth.

42. *Dialogue,* p. 197.
43. *Dialogue,* p. 198. By definition of *G*,

$$\frac{FE}{GE} = \frac{BI}{AI}.$$

Adding 1 to both sides gives

$$\frac{FE}{GE} + \frac{EG}{EG} = \frac{BI}{AI} + \frac{AI}{AI}.$$

Multiplying by 1 gives

$$\frac{FG}{EG} \cdot \frac{EG}{EG} = \frac{AB}{AI} \cdot \frac{AB}{AB};$$

but since $AI \cdot AB = C^2$ and $FG \cdot GE = GH^2$ (i.e., *C* and *GH* are mean proportionals to the other values), this yields

$$\frac{GH^2}{EG^2} = \frac{AB^2}{C^2} \quad \text{or} \quad \frac{GH}{EG} = \frac{AB}{C}.$$

44. *Dialogue,* p. 199.
45. *Dialogue,* p. 203.
46. Only the fourth proposition of the preliminary *De Vi Centrifuga,* where Huygens computes the radius of the circle for which the centrifugal force of the earth would equal gravity and compares it with Snell's value for the actual radius (see note 17), deals with the spinning earth; all others are about tension on cords and balls moving on surfaces of revolution in 1 second.
47. *Dialogue,* p. 449.
48. *Dialogue,* p. 216.
49. See their introduction to *De Vi Centrifuga, OH* 16:240–1.
50. *Dialogue,* pp. 217–18.
51. *Dialogue,* p. 211.
52. "Galileus deceptus" (*OH* 16:251, n. 3). In his *From Galileo to Newton,* pp. 51–3, A. R. Hall has pointed to the weaknesses in Galileo's treatment of centrifugal force but David Hill argues that Galileo most likely knew he was giving an erroneous argument but persisted for rhetorical reasons. In a reply to Hill, Drake goes further and argues that Galileo was in some sense correct, because a body on the earth cannot ever behave as the standard sling model suggests.
53. See particularly Propositions 16 and 17 of *Livre second: Du mouvement des corps,* pp. 137–46 in *Harmonie universelle.*
54. "Cecy estant posé, il faut examiner par nombres ce que Galilee a voulu demonstrer par lignes, à scavoir que l'espace que doit faire le corps ietté par le mouuement iournalier de la terre pour se reunir à sa surface est si petit, qu'il n'a que trop de temps pour y arriver" (*Harmonie universelle,* p. 142).
55. Except for errors in calculation, Mersenne's ratio was as good as Huygens's value; for details, see MacLachlan. Mersenne discovered by numerical example the principle that Huygens was later to use, namely that the parabola and circle maintain a constant ratio between their values near the point of contact. Fitting the two curves, as Huygens did, requires that the ratio be made equal to 1.

56. In speaking of a variation in gravity, Mersenne was thinking of very ligh⸀ objects whose resistance to air is so weak that they cannot fall 12 feet in 1 second as do most objects. Of course, the resistance of air is not a factor in Huygens's analysis, which depends only on adjusting the speed of rotation or the radius of the circle.

57. "Nous auons donc monstré, qu'il n'est pas veritable qu'encore l'on augmentast le mouuement par la tangente, & que l'on diminuast celuy qui se fait par la secante, que le chemin que le poids deuroit faire pour arriuer à la circonference, seroit si petit, que quelque temps qu'il y eust, it seroit toujours trop suffisant" (*Harmonie universelle,* p. 145).

58. *Harmonie universelle* is the only work in which both topics are discussed, the pendulum experiment on pp. 131–7 and centrifugal tendency on pp. 137–46.

59. Both works appear on the list of Huygens's library holdings compiled at his death; see the list appended to *OH* 22 (sep. pag.), p. 1, No. 21, and p. 5, No. 25.

60. Galileo, *Two New Sciences,* pp. 162–4 for the argument, p. 163 for the quotation.

61. *OH* 16:301.

62. "Libera per vacuum posui vestigia princeps" (Horace, *Epistles* 1.19.21; quoted in *OH* 16:302).

4. ACCELERATED MOTION: CURVILINEAR FALL

1. The placement and numbering of these propositions as presented in *OH* 16:295–301 differs from those in the first draft; see *OH* 16:311, n. 3.

2. Galileo, *Two New Sciences,* pp. 210–13; see particularly his scholium to Proposition 36, Theorem 22.

3. *OH* 17:125–37. The figure from Huygens's last proposition on fall also appears on p. 13 of the manuscript containing *De Vi Centrifuga,* between the first draft and its revision, showing how very closely the two works were related temporally (*Hug.* 26, f. 10r).

4. Two arguments regarding fall through all degrees of speed occur on ff. 80r–v of *Hug.* 10; then the treatise on inclined planes begins in the middle of f. 80v. The editors chose to reprint the beginning paragraphs in *OH* 17:278–81 rather than with the treatise (*OH* 17:125–37), although the theorems form a logical prelude to the main work and there is no demarcation in the manuscript between the two sections.

5. *OH* 17:125–8; cf. Galileo, *Two New Sciences,* Theorem 2, pp. 166–8.

6. *OH* 17:128–30; cf. Galileo, *Two New Sciences,* Theorem 1, pp. 165–6.

7. *OH* 17:130–2; cf. Galileo, *Two New Sciences,* Postulate, p. 162.

8. *OH* 17:132–3; cf. Galileo, *Two New Sciences,* Theorem 10, pp. 185–6.

9. *OH* 17:133–6.

10. They form Propositions 1 to 11, which are followed by a series of theorems leading to the proof of the isochronism of the cycloid (*OH* 18:127–49).

11. "Spatia super diversis planis inclinatis eodem tempore ab eodem mobili peracta, sunt inter sese ut potentiae quibus in unoquoque plano sustineri possunt" (*OH* 17:136).

In the 1655 edition of the *Discorsi*, Viviani adds a proof of the "postulate," which Galileo had dictated to him in 1638–9 and which uses the same mechanical principle as Huygens's proof but is stated in terms of impetus; see Galileo, *Two New Sciences*, pp. 171–5.

12. *OH* 17:137. It is unclear whether by "lengths" Huygens intends the sum of the lengths of the various planes or the length of the bottom-most planes extended to the initial altitude. Since he states the relationship in terms of proportions, either is correct.

13. Proposition 21 of Part 2 of the *Horologium Oscillatorium* says that given two sets of contiguous planes, where each plane in the first set has the same altitude as its mate in the second set but has a shallower pitch to its incline than the other, the time of fall through the first set will be shorter than that through the second set. Huygens then generalizes this to two curvilinear segments cut by parallel horizontal lines. The curve whose tangent at any point is less steeply pitched than the tangent of the other curve at every comparable level is the curve of quicker descent (*OH* 18:165–7).

14. "Tempora vibrationum pendulorum in subduplicata ratione longitudinum. Tempora circulationum horizontalium sunt in subduplicata ratione axium conorum quos describunt" (*OH* 16:319).

15. "Quaeritur tempus per quadratem Circumferentiae circuli quod dubito an inveniri possit" (*OH* 18:374). In fact, this particular attempt might postdate the isochronism proof. Its proof and diagram are analogous to those used in the infinitesimal situation, and it occupies the verso of the list of lemmas necessary for that derivation. However, a very similar diagram, unaccompanied by any derivation to clarify its meaning, appears in the middle of the first draft of *De Vi Centrifuga* (*Hug.* 26, f. 6v), which was definitely completed by Nov. 15. Thus, Huygens attempted at least once before Dec. 1 to solve the problem for the quarter arc, and with a method akin to that used successfully in the infinitesimal case and continually unsuccessfully in the larger case. See *OH* 18:376–87 for further attempts to solve the problem for the quarter arc made by Huygens in 1669 and 1693. In all cases, he is blocked by his inability to perform the geometric equivalent to integrating the function $\sqrt{a^5/(a^2x - x^3)}$.

16. "Quaeritur quam rationem habeat tempus minimae oscillationis penduli ad tempus casus perpendicularis ex penduli altitudine" (*OH* 16:392).

17. For derivations of this proof in modern notation, see Dugas, pp. 302–5, and Whiteside's commentary in *Mathematical Papers of Newton* 3 (1969): 392–3.

18. The complete transcription of his proof fills *OH* 16:392–7; originally *Hug.* 26, f. 72r. The editorial commentary was provided primarily by Dijksterhuis, who repeated his arguments in "Ontdekking."

19. *TE* is perpendicular to the infinitesimal at *E*; *BE* is perpendicular to the infinitesimal at *B*. Therefore, the right triangles are similar.

20. Let *e* be the speed at point *E*, because distance is proportional to speed squared (as well as time squared),

$$\frac{AB}{AZ} = \frac{e^2}{v^2}; \quad \text{but} \quad \frac{AB}{AZ} = \frac{BD^2}{Z\Sigma^2}$$

by the definition of the parabola. Therefore, e/v equals $BD/Z\Sigma$ and BD can be considered e's value in any statement involving proportions, because v and $Z\Sigma$ are constants.

21. "Sine quibus motus aequabilis in cava cycloide inveniri non poterat" (*Hug.* 26, f. 73r; *OH* 16:398–400).

22. See the earlier section of the derivation and *OH* 16:398.

23. *OH* 16:398.

24. *OH* 16:399. The lettering is changed to agree with that in Fig. 4.3. Huygens's proposition follows easily from the definition of the parabolas. In parabola $AD\Sigma$, $bd^2/BD^2 = Ab/AB$, and in parabola $ZE\aleph$, $be^2/BE^2 = Zb/ZB$. Thus,

$$\frac{(be \cdot bd)^2}{(BE \cdot BD)^2} = \frac{Ab \cdot Zb}{AB \cdot ZB} \quad \text{or} \quad \frac{bw^2}{BW^2},$$

because by mean proportions on the circle, $AB \cdot ZB = BW^2$ and $Ab \cdot Zb = bw^2$. Note that it is not necessary for the two parabolas to be the same size, although they are in Huygens's proof.

25. *OH* 16:399. The lettering is changed to agree with that in Fig. 4.3.

26. The reconstruction by the editors (Dijksterhuis) of this proposition is more involved, since they introduce times of fall into the discussion (*OH* 16:399, n. 6). Their argument, like mine, is based on an appeal to infinitesimals similar to that made by Huygens at the beginning of his derivation.

27. *OH* 16:401–3. Again, I have changed the lettering. Huygens seems to have had no regard for consistency in labeling his diagrams. Indeed, it is so rare to see two diagrams with the same lettering that I am willing to claim that the work done here preceded by a very short time, perhaps only hours, a much more refined proof (*OH* 16:404), since the drawings are exactly the same except that those parts of the diagram used in the first proof but not needed in the revision are eliminated.

28. By the definition of points on the curve LXN,

$$\frac{CN}{Z\Sigma} = \frac{BG \cdot Z\Sigma}{CQ \cdot CQ},$$

because for the midpoint C, CQ is the corresponding value on both parabola $ZE\aleph$ and parabola $AD\Sigma$ and, in addition, $BF = Z\Sigma$. However, since Σ and Q are on a parabola and C is the midpoint of AZ,

$$\frac{Z\Sigma^2}{CQ^2} = \frac{AZ}{AC} = 2.$$

Thus, the original equation, which can be rewritten as $CN = BG(Z\Sigma^2/CQ^2)$, becomes $CN = 2BG$, or $2b$, since $BG = TZ = b$. Because C is the midpoint of AZ, CI is $c/2$. Thus, $C\Pi$ follows from $CI/C\Pi = \text{arc } AIZ/AZ$, which gives $C\Pi = cq/2p$, where p/q is equivalent to $\pi/2$. Because $Z\Sigma$ is on a parabola with a *latus rectum* defined to be $2b$, $Z\Sigma^2 = (2b)(c)$ and thus $Z\Sigma = \sqrt{2bc}$. Finally, because $ZU = CI$,

$$\frac{C\Pi}{CO} = \frac{ZU}{Z\Sigma} \quad \text{makes} \quad CO = \frac{Z\Sigma \cdot C\Pi}{CI} = \frac{(\sqrt{2bc})(cq/2p)}{c/2} = \frac{q\sqrt{2bc}}{p}.$$

29. "Hoc autem Cycloidi convenire inveni ex cognita tangentis ducendae ratione" (*OH* 16:397).
30. See *OH* 16:400 for the auxiliary proposition.
31. Both Roberval and Descartes had devised methods for drawing tangents to a cycloid; for a description of their work, see Foà. Huygens developed his own variant of their techniques (*OH* 14:374–5).
32. In fact, they assume that Huygens knew or guessed the cycloid was the correct curve and needed only to derive the tangential relation. Many other commentators, including Dugas (p. 305), appear to accept this conclusion.
33. For an example of a later proof that uses this fact, see *OH* 16:404–5. If the curve is a cycloid, with *MPA* its generating circle, then

$$\frac{DB}{CB} = \frac{MP}{CP} = \frac{AP}{AC}.$$

Therefore, the defining relationship

$$\frac{DB}{CB} = \frac{CE}{CF}$$

can be written

$$\frac{CE}{CF} = \frac{AP}{AC}$$

or, squaring,

$$\frac{CE^2}{CF^2} = \frac{AP^2}{AC^2}.$$

In the circle, $AP^2 = MA \cdot AC$, so that inversion gives

$$\frac{CF^2}{CE^2} = \frac{AC^2}{MA \cdot AC}$$

and thus

$$CF^2 = \frac{CE^2 \cdot AC}{MA}.$$

That is, *F* lies on a parabola with a *latus rectum* CE^2/MA. Therefore, for the ordinate value *MA*, the parabola yields the value *CE*.
34. He would have $CF^2/AP^2 = p(AC)/(MA \cdot AC)$, or p/MA, and so *CF* would be proportional to *AP*.
35. See Figure 4.2, bottom.
36. *OH* 16:401–3 contains a fragmentary proof, which I have included above as one of the auxiliary propositions. However, its drawing clearly contains the addition of the generating circle of the cycloid, although in other respects it is similar to the main text. The same sketch – even the same lettering, which is highly unusual with Huygens (see note 27) – is repeated in another proof (*OH* 16:404–5), except that the infinite spaces are missing since the derivation has no need for them. Still another proof (*OH* 16:405–12), labeled by Huygens as "demonstratio melior huc tandem redacta," uses an approach that

deals only with the generating circle and tangents to the cycloid in a manner similar to the proof in the *Horologium Oscillatorium* (*OH* 18:171–87). Huygens does continue to refine this proof; see *OH* 17:138–41, for example.

37. "Magna nec ingenijs investigata priorum" (Ovid, *Metamorphoses* 15.146; quoted in *OH* 16:406).

38. Ariotti, p. 397. Koyré is another who claims that Brouncker had the cycloid first (p. 226, n. 23). Brouncker did attempt to prove the isochronism of the cycloid – after having been told of Huygens's discovery by Moray; for the evidence, see Ziggelaar, "Tautochronisme."

39. Ziggelaar argues that Pardies probably knew of Huygens's result from Auzout, who was trying to prove the result himself ("Tautochronisme," p. 32). Costabel argues that, although Huygens's claim of originality is true, the proofs presented by Brouncker and Pardies have value because those mathematicians – and Newton – approached the problem in significantly different ways. In particular, he claims, Huygens had to rework his own proof before he could realize the general case for isochronism ("Isochronisme," pp. 15–16). Ziggelaar goes further, hinting that Huygens's generalization of the requirement for isochronism came directly from seeing Pardies work ("Aux origines").

40. "Mais a mon avis il ne s'agit pas en cecy de l'honneur de Milord Brouncker, puisque ce n'est pas grande chose pour luy d'avoir fait la demonstration d'une proposition desia trouvée, ce que d'autres ont fait aussi comme le P. Pardies, de qui je ne puis m'empescher icy, de vous dire, que je regrette extresmement la perte. Le principal, et ce qu'il y a de plus difficile dans ces choses de Geometrie, c'est de les trouver comme scavent tres bien ceux qui s'en meslent" (*OH* 7:314).

41. See *OH* 16:320–1 for the geometric argument; it is accompanied in the manuscript by an algebraic version that continues the notation of a similar algebraic derivation at the end of his isochronism proof, thus supporting my chronology of events.

42. Huygens's basic formula is $d = 2\pi^2 L$, where d is the distance fallen in the same time t as one rotation (double oscillation) of a conical (cycloidal) pendulum of height (length) L. If the times-squared law is applied, $t^2/(1 \text{ sec})^2 = d/(g/2)$, whence, if $2\pi^2 L$ is substituted for d, $t^2 = 4\pi^2 L/g$ and $t = 2\pi\sqrt{L/g}$.

Both Huygens's editors (*OH* 17:246) and Koyré (pp. 233–4) feel that Huygens did not derive the formula until his work on the cycloid, although his conical pendulum predates the cycloidal one and the formula is as explicitly stated for the conical as for the cycloidal pendulum. Obviously, Huygens never expressed the modern formula for either pendulum.

43. *Hug.* 26, f. 72v; not reproduced in *OH*.

44. What might be Huygens's original calculation (done sometime after Nov. 15) appears on f. 130v of *Hug.* 10. In that derivation he does not specifically calculate the remainder for the diameter of the circle circuited in 1 second. Perhaps, reflecting this laxness, the value for the diameter recorded in *De Vi Centrifuga* (*Hug.* 26, f. 15v) has an inaccurate, not to mention awkwardly expressed, remainder: $d = 1.5\frac{3}{5}$ feet, $r = .7\frac{8}{10}$ feet (i.e., 1.56 feet and 0.78 feet). In this new calculation, which appears at the bottom of *Hug.* 26, f. 72v,

he finds the diameter to be $(15.6 \cdot 7 \cdot 7)/(22 \cdot 22)$, or 1.5 feet with a remainder of $^{384}\!/_{484}$, which this time he converts to inches by multiplying by $^{12}\!/_{10}$, getting 4608/4840. Dividing both 1.5 feet (18 inches) and the remainder by 2, Huygens gets $r = 9^9\!/_{20}$ inches.

45. The calculation is written across work previously done at the top of the page. Taking $L = 9\frac{1}{2}$ inches, or $2L = 19$ inches, Huygens computes $(19 \cdot 355 \cdot 355) \div (113 \cdot 113)$, which he rounds off to $187\frac{1}{2}$. Dividing by 12 gives him 15 feet $7\frac{1}{2}$ inches.

46. The passage in question originally began: "Quum experientia doceat..." Huygens crosses out "experientia" and replaces it with "calculus" and a dagger identifying a marginal note: "Imo calculus, ut postea inveni, postquam propositio casus perpendicularis ad casum per cycloidem sive penduli vibrationem innotuit" (*OH* 16:278). In the margin of the manuscript there is also a recomputation of the radius of the seconds-circle using $\pi = {}^{22}\!/_7$ (vs. $^{355}\!/_{113}$) and 15 feet $7\frac{1}{2}$ inches for the constant of acceleration; i.e., he really has computed it for this passage. Again, his editors report the two values of the constant of gravitational acceleration in such a way that nothing significant appears to have occurred between Huygens's recording of $15^6\!/_{10}$ feet and its replacement 15 feet $7\frac{1}{2}$ inches.

47. Misled by Huygens's editors into thinking that the value was constant (i.e., that the radius was always $9\frac{1}{2}$ inches), Costabel does try to provide an explanation based solely on the interpretation of the word "imo" in Huygens's marginal note ("Isochronisme," pp. 8–12).

48. "Je ne trouve pas qu'il soit necessaire d'egaler le mouvement du pendule par les portions de Cycloide pour determiner cette mesure, mais qu'il suffit de le faire mouvoir par des vibrations fort petites, les quelles gardent assez pres l'egalité des temps, et chercher ainsi quelle longueur il faut pour marquer par exemple, une demie seconde[a] par le moyen d'une horologe qui soit desia en train de bien aller, et ajustee avec la Cycloide.... a) J'ay trouvé cette longueur de $9\frac{1}{2}$ pouces bien exactement de nos pieds de Rhynlande, a scavoir depuis le point de suspension jusqu'au centre de la boule, que je prenois de diverse grandeur et differente matiere" (*OH* 3:438, 440, Dec. 30, 1661). Huygens also describes this method in the *Horologium Oscillatorium,* Part 4, Proposition 25 (*OH* 18:349–51).

49. In that experiment (see Chapter 3 for a description) the constant of gravitational acceleration is $105.5(60/45)^2$, or 187.55...inches, which when truncated to tenths is 187.5 inches. Dividing by 12 gives 15.625 feet, which truncates to 15.6 feet. Actually, Huygens does not arbitrarily truncate; rather he appears to carry his computations to the tenths position and then guesses the remainder, which if around one-half or less is then dropped.

50. Even the identification of $9\frac{1}{2}$ inches with 15 feet $7\frac{1}{2}$ inches is dependent on a rounding to tenths, since 15 feet $7\frac{1}{2}$ inches yields a length for the pendulum of 9.49 inches.

51. According to his editors, the Rhenish foot used by Huygens equals 0.3139 meter, making 15 feet $7\frac{1}{2}$ inches the equivalent of 980.9375 cm/sec^2. They also state, in what can only be viewed as partisan enthusiasm, that $9\frac{1}{2}$ Rhenish inches "coincides exactly" with the true length of the seconds-pendulum

for the latitude of The Hague (*OH* 3:440, n. 13). Of course, it does vary with latitude, as even Huygens later knew.

In its latest bid for accuracy, science has returned to using a clock to determine the basic unit of length, just as Huygens wanted with his horological foot. The meter is currently defined by the distance orange-red light from krypton-86 travels in 1 second, as measured by the atomic clock of vibrating cesium-133.

52. Actually, he wished to use as the standard the pendulum that made a single oscillation (one full swing without returning) in 1 second. He suggested defining its length (38 Rhenish inches) to be 3 horological feet; see his comments in the *Horologium Oscillatorium* (*OH* 18:349–53).

5. EVOLUTES

1. "Mais je vous puis assurer, que tant s'en faut que l'addition du poids fasse haster le pendule, que au contraire elle le rend tant soit peu plus lent, luy donnant un mouuement plus large, tout ainsi que du simple pendule les coups qui s'eloignent le plus de la perpendiculaire sont plus lents que les autres. et mesme pour remedier a ce defaut contraire a celuy que vous craigniez je suspendois du commencement le pendule entre deux platines courbes comme *AB*, *CD*, que l'experience m'apprit de quelle maniere et combien je devois plier, pour esgaler entre eux les coups des plus larges jusqu'aux plus menus. Et je me souviens d'avoir si bien adjustè deux Horloges de cette façon, qu'en trois jours ils n'y eust jamais entre elles la difference d'autant de secondes: quoyque cependant j'en changeasse souvent les poids, les rendant plus ou moins pesants. Toute fois par apres parce que je trouuois que avec ces platines la moindre inclination de l'horloge alteroit la longueur du pendule, je les ay cassees faisant deuenir en mesme temps les vibrations du pendule plus estroites, par le moyen des roues" (*OH* 2:271, Nov. 1, 1658).

2. See *OH* 17:14–20 for pictures and manuscripts relating to his clocks of 1657.

3. *OH* 17:17.

4. *OH* 2:272. See Galileo, *Dialogue,* pp. 230, 449–51, where he also claims that the quarter arc of a circle is the curve of quickest descent, outlining the ideas handled more rigorously in *Two New Sciences*. See also Mersenne, *Harmonie universelle,* p. 135, and *Cogitata,* Proposition 16, p. 45, in which he states that the time is proportional to the square root of the length.

5. See the discussion in Chapter 3 and *OH* 16:301.

6. For examples, Dijksterhuis's claim in "Ontdekking," pp. 193–4; Landes, p. 119.

7. For his relativistic argument, see *OH* 16:263–7.

8. *OH* 17:145. The editors think that this derivation is close to the original discovery proof (because Huygens does not apply the general theorem for evolutes), but they conjecture that it is a copy of the original (with perhaps some changes) because it appears on a sheet that otherwise contains neat drawings for a marine clock not invented for another couple of years. Yet those figures were originally done in pencil and then inked over, which gives them a less finished standing than the editors suggest. More to the point, the curved plate

for the close-up of the pendulum's mount just happens to lie along the evolute of the proof. From what I know of Huygens's habits, I find it hard to believe that he would bother to recopy this proof, taking care to wrap the evolute of its diagram along the plates of the figure supposedly already on the page, and then flip the paper and begin a slightly improved analytic version (*OH* 17:143–4). He would have begun immediately to search for an analytic proof suitable for publication. In addition, the diagram for this proof has point *W* of the evolute on the wrong side of the axis; i.e., when Huygens drew this figure he obviously did not know *W* lay on a cycloid that belongs completely to the left of the axis. In his improved derivation on the overside, the normals do intersect in the proper position. Would he really have recopied an incorrect figure? I conclude that this is the original discovery proof and that the drawings, neat as they are, were added later, perhaps as test sketches for those in the *Horologium Oscillatorium,* which they greatly resemble.

9. Point *W* cannot really be on the evolute, since it would follow that both *FW* and *VW* are tangents to the evolute at that same point. However, the separate points on the evolute at which the two normals are tangent are very close to each other and, in addition, each is close to *W*. Thus, as *FV* approaches 0, this separation also approaches 0. Huygens argues similarly when he comes to formulate the general case.

10. Two tangents to a circle form an isosceles triangle with the chord joining their points of tangency as base; see Euclid, *Elements,* Book 4, Proposition 12. In this case, angle *OCK* equals angle *OKC*. Huygens used this construction extensively in his *De Circuli Magnitudine Inventa* (*OH* 12:121–81).

11. A right triangle subtends a semicircumference and can thus be considered to be inscribed in a circle where its base is the diameter of the circle; see Euclid, *Elements,* Book 4, Proposition 5.

12. Pascal issued the challenge under the pseudonym Dettonville (d'Ettonville). The first set of problems appeared in June 1658 and was sent to Huygens by Boulliau; see *OH* 2:186–9, which includes the text of the challenge.

13. "Mais entre tous les ecrits qu'on a recus de cette sorte, il n'y a rien de plus beau que ce qui a ete envoye par M. Wren; car outre la belle maniere qu'il donne de mesurer le plan de la Roulette, il a donne la comparaison de la ligne courbe meme et de ses parties avec la ligne droite. Sa proposition est que la ligne de la Roulette est quadruple de son axe, dont il a envoye l'enonciation sans demonstration" (reprinted in *Oeuvres de Blaise Pascal* 8:204). The Latin edition appeared under the title *Historia Trochoidis sive cycloidis* at the same time, namely October 1658.

14. Huygens's copy was sent by Boulliau on Jan. 3, 1659 (*OH* 2:308–9).

15. "Uyt de historia Trochoidis, van Pascal en Roberval gemaeckt, heb ic Wrens inventie verstaen, quae magni facienda et mirifice placuit. Is het beste datter van die linie gevonden is. Want ic distinguere de moeyelycke van de elegantia. Ick vond de demonstratie en generaelyck &c." (*OH* 2:330, summary dated Jan. 31, 1659). Only the summary exists. Huygens also announced his discovery to Sluse on Jan. 14 (*OH* 2:312–13) and to Carcavy via Boulliau on Jan. 16 (*OH* 2:315).

16. Huygens's own proof is dated Jan. 11 (*OH* 14:363–7). At first glance his proof looks very sloppy and complicated. This impression is reinforced by Sloth, who indiscriminately repeats in modern notation the entire text presented in *OH*, including even an erroneous claim made by Huygens but never actually used. A closer examination shows that the passage consists of two proofs: The first is Wren's result for the whole cycloid; the second is the proof that I outline here with excess, unused statements removed. The diagram, used for both proofs, is cluttered with elements not necessary for the second.

17. Archimedes, *De sphaera et cylindro,* Book 1, Proposition 22, in *Opera Omnia* 1:99–101.

18. See *OH* 14:364 for the lemma.

19. In contrast, Wren's proof is much more straightforward and plodding, although it is also much more amenable to strict application of modern limit procedures. Of course, his proof was undoubtedly tidied for publication, whereas Huygens's is the original discovery proof and was never prepared for publication, since his published version uses the theory of evolutes.

20. *OH* 2:360. Wallis was Huygens's primary British correspondent on mathematical matters; he was also Wren's scientific spokesman. Wallis's reply of Feb. 28, 1659, contains a great many complaints regarding Pascal's handling of the competition, which presaged the animosity between British and Continental mathematicians that flared during the priority squabble over the calculus. His *Tractatus duo: Prior, de cycloide et corporibus inde genetis. Posterior, epistoris; in qua agitur, de cissoide, et corporibus inde genetis...* (1659; reprinted in his *Opera Mathematica* 1:489–542) was his answer to Pascal's *Histoire*. The letter of the second treatise, by the way, was addressed to Huygens.

21. Wren's proof was finally published by Wallis in *Tractatus duo* (*Opera Mathematica* 1:533–7). Wallis sent Huygens a copy on Dec. 4, 1659, but it was not received until Mar. 20, 1660 (*OH* 2:518–20 and 3:58).

22. *OH* 14:404.

23. The cycloid is defined mechanically as the path traced by the point D, situated on the circle MDC, as the circle rolls along the baseline FK. At D the circle has rolled the distance FC, which equals arc DC on the circle.

24. Indeed, even if Huygens did not discover the shape of FBA by the process that I have reconstructed, he certainly would have used his formula $FB = 2FL$ to find the full length of the pendulum cord. He would then have the fact that FBA equals EA, or twice the diameter EK. But what curve has a length equal to twice the diameter of a circle? The semicycloid, because the full arc of the cycloid is four times the diameter.

25. For the astronomy, *OH* 15:534–50; for the mathematics, *OH* 17:98–9; for the instructions, *OH* 17:100. *Hug.* 10, f. 126r (not reproduced in *OH*), contains many computations of L for various combinations of gears. Since the time of oscillation is as the square root of the length of the pendulum, the length L will be as the square of the inverse of the number of oscillations n. Thus, $L/9.5 = (3,600/n)^2$, or $Ln^2 = 123,120,000$. Note that by this time Huygens had definitely accepted 9½ inches as the length of the pendulum that beats in 1 second. Koyré (p. 234) interprets the instructions as proof that Huygens did an experiment with the simple pendulum to determine the constant of

gravitational acceleration, citing the example given by Huygens in his instructions (which is one of the thirteen combinations on f. 126r) as an experimental value.

26. "Occupatum me habet novum inventum quod hisce diebus excogitavi ad horologium meum exactius etiamnum efficiendum quam fuit hactenus. Scis puto adhibuisse me in automatis istis lamellas binas incurvatas ut *AB*, *AC*, inter quas pendulum suspensum movebatur; idque propterea factum ut omnes penduli vibrationes aequali tempore redirent, quae alioqui non omnino ἰσόχρονοι erant ut in libello meo indicavi. Quod igitur nunquam me inventurum speraveram nunc denique reperi, veram nimirum figuram curvarum *AB*, *AC*, quae efficiat ut oscillationes omnes accuratissime exaequentur. Eam ratione geometrica determinavi, ipsosque artifices docebo ut nullo negotio curvam illam lineam describant. Subtilissimo Heuratio non displicebit opinor haec inventio; nam mihi quidem omnium felicissima videtur in quas unquam inciderim" (*OH* 2:522).

27. *Hug.* 26, f. 75r; not reproduced in *OH*. The proof given in the *Horologium Oscillatorium* is synthetic, starting with the semicubical parabola as given and determining that the parabola is its involute (*OH* 18:207–9).

28. Given parabola $y^2 = rx$, where r is the *latus rectum,* the subnormal ($y\,dy/dx$) to any point x would be $r/2$.

29. *FLE* and *BGE* are right triangles with opposite angles, *BEG* and *FEL*, equal and thus the triangles are similar.

30. From the similarity of triangles *BGK* and *EGB*, $BG/KG = EG/BG$, or $BG^2 = EG \cdot KG$. By the definition of the parabola, $BG^2 = rGH$, and $r = 2EG$ (see note 28). Thus, BG^2 also equals $2EG \cdot GH$, and KG must equal $2GH$.

31. A semicubical parabola is a curve of the form $aY^2 = X^3$; i.e., Y is as the three-halves (semicubical) power of X. Since in classical style the dimensions on either side of the equation must be the same (in this case three), a is considered to be a line of dimension 1. In the original parabola, $y^2 = rx$, the missing dimension was added to x's side of the equation and thus the constant (*latus rectum*) appears on different sides of the "standard form" for these curves.

32. The earliest extant drawing of a semicubical regulator for a rotating pendulum dates from 1663–4 (*OH* 17:153). Huygens's editors push the date of actual construction back to 1667–8, when Huygens writes to his brother Lodewijk about a new conical clock (*OH* 16:243, n. 9, and 18:437). In a letter to his father during a priority squabble with Robert Hooke, Huygens claims that he invented the paraboloidal-pendulum clock at the same time as the cycloidal one – in 1658 (*sic*); see *OH* 7:390–2 (Aug. 17, 1674) and his reiteration (*OH* 7:431, Mar. 29, 1675). However, it is unclear whether Huygens refers to the discovery of the principles behind the clock or to the actual construction of an appropriate mechanism.

33. In Part 5 of the *Horologium Oscillatorium* Huygens describes a clock of this design but begins by referring to the difficulty of its construction (*OH* 18:361).

34. For the application to the ellipse, see *OH* 14:387–90. The editors do not include all of Huygens's calculations; cf. *Hug.* 10, ff. 91r–93v. Folio 94v (an improved proof of the isochronism of the cycloid) is dated Dec. 15, thus delimiting the period during which Huygens developed the theory of evolutes.

35. *OH* 18:225-9.
36. $HN/HL = (HN/HF) \cdot (HF/HL)$, which, by the similarity of triangles, is the same as $(ds/dx) \cdot (ds/dx)$, or $(ds/dx)^2$.
37. $BG/MG = BG/(BG - BM)$ and thus $MG/BG = (BG - BM)/BG$, or $1 - BM/BG$. Substituting the algebraic values gives

$$\frac{(ds/dx)^2 + y(d^2y/dx^2)}{(ds/dx)^2} = 1 - \frac{y(ds/dx)}{BG},$$

and subtracting 1 from each side of the equation yields

$$\frac{y(d^2y/dx^2)}{(ds/dx)^2} = \frac{-y(ds/dx)}{BG}$$

and thus

$$BG = \frac{-(ds/dx)^3}{d^2y/dx^2}.$$

Because of its geometric meaning, the radius is taken as the positive of this value.
38. Bruins takes Huygens to task for not realizing that polygonal paths are counterexamples of his claims.
39. Since $KL + LN = KM + MN$, then $MN - KL = LN - KM$ (∗). For each point on the axis, such as K and L, draw a line perpendicular to the axis and equal to the subnormal of the curve for that point; e.g., $KT = KM$, $LV = LN$. The locus of points T, V, etc., defines the auxiliary curve. The ratio of the difference between two values on the curve, $LV - KT$ ($= LN - KM$), to the vertical difference KL, when KL is assumed infinitesimal, will be the same as the slope of the tangent at T. Thus, assuming the tangent can be found, $(LN - KM)/KL$ exists, which by the equation (∗) is the same as $(MN - KL)/KL$. Adding 1 gives $KL/KL + (MN - KL)/KL$, which is MN/KL, the reciprocal of the value being sought. See Huygens's argument in the *Horologium Oscillatorium* (*OH* 18:231-3).
40. In the parabola $y^2 = rx$, $y\,dy/dx$ equals $r/2$; i.e., the curve described by the subnormal is a vertical line. In the ellipse $ay^2 + bx^2 = c$, $y\,dy/dx = -bx/a$, yielding a line of slope $-b/a$. In the hyperbola $ay^2 - bx^2 = c$, $y\,dy/dx = bx/a$, a line of slope b/a. For the semicubical parabola $y^3 = ax^2$, the result can be written as $27(y\,dy/dx)^3 = 8a^2x$, which is the form of a cubical parabola. See *OH* 18:233-5 for Huygens's presentation in the *Horologium Oscillatorium*.
41. In the parabola, $LN = r/2 = KM$ and therefore $(LN - KM)/KL = 0$. Adding 1 (see note 39) gives $MN/KL = 1$. In modern notation, $d(y\,dy/dx)/dx = d(r/2)/dx = 0$, and thus $1 + d(y\,dy/dx)/dx = 1$.
42. Huygens's derivation is algebraic, not geometric; see note 34. The auxiliary curve is $-bx/a$, and thus the change in that curve to the change in x $[d(y\,dy/dx)/dx]$ is $-b/a$. Adding 1 gives $MN/KL = (a-b)/a$ and, therefore, $KL/MN = a/(a-b)$, which is a constant.
43. Since K and L are infinitesimally close, HN/HL can be considered equal to HM/HK. In modern notation, $HN/HL = (ds/dx)^2$, and thus

$$\frac{HN}{HL} = 1 + \frac{b^2x^2}{a^2y^2},$$

or, removing the dependent variable,

$$\frac{HN}{HL} = \frac{a^3b - 4abx^2 + 4b^2x^2}{a^3b - 4abx^2}.$$

Huygens achieves the same algebraic result by applying Proposition 36 of Apollonius, *Conics*, Book 1. Therefore,

$$\frac{BG}{MG} = \frac{KL}{MN} \cdot \frac{HN}{HL} = \left(\frac{a}{a-b}\right)\left(\frac{a^3b - 4abx^2 + 4b^2x^2}{a^3b - 4abx^2}\right)$$

$$= \frac{a^3 - 4ax^2 + 4bx^2}{(a-b)(a^2 - 4x^2)} = \frac{a^3 - 4ax^2 + 4bx^2}{a^3 - 4ax^2 - a^2b + 4bx^2}.$$

44. Huygens's calculations take up seven manuscript pages, not all of which are reproduced in *OH* 14:391–6. In general, $KL/MN = a/(a+b)$, and thus

$$\frac{BG}{MG} = \frac{4ax^2 + 4bx^2 - a^3}{(a+b)(4x^2 - a^2)}.$$

When a equals b, the evolute can be written as $(y^2 - a^2 - x^2)^3 - 27a^2y^2x^2 = 0$.
45. *Theoremata de quadratura hyperboles, ellipsis et circuli ex dato portionem gravitatis centro. . .* (Leiden, 1651); reprinted with facing-page French translation in *OH* 11:281–337.
46. *OH* 14:397–8.
47. *OH* 14:399–403.
48. *OH* 18:191–9, particularly Proposition 4.
49. *OH* 17:120–3.
50. *OH* 16:405–12.
51. *OH* 17:146–7. The material appears in the *Horologium Oscillatorium,* Part 3 (*OH* 18:233–41). Huygens is concerned with curves of the general form $y^m = ax^n$.
52. His first attempt occurred immediately after he discovered the isochronism of the cycloid; see *OH* 16:385–91.
53. "Vous verrez dans une seconde edition du dit horologe, plus ample que la premiere, une belle invention que depuis peu j'y ay adjouste pour la derniere perfection. C'est le fruit principal que l'on pouuoit esperer de la science de motu accelerato, que Galilée a l'honneur d'avoir traictée premier: et je m'assure que les Geometres estimeront infiniment plus cette addition que tout le reste de cet automate" (*OH* 3:13, Jan. 22, 1660).

6. CURVATURE

1. For example, Haas begins his history of curvature with a brief description of Huygens's work on evolutes. Coolidge uncovers earlier discussions of curvature in Oresme, Kepler, and others, then discusses Huygens and evolutes in his sketchy history.
2. Actually, in modern mathematics, "curvature" is defined first as the rate of change of the angle of the tangent with respect to the arc length, and then the radius of curvature is introduced, as the positive reciprocal of the curvature. Thus, for a straight line, the curvature is zero and the reciprocal is undefined or infinite according to one's mathematical taste.

3. See his derivations, pp. 289–93.

4. Heath, in fact, labels the section in which he summarizes the applicable Apollonian propositions as "propositions leading immediately to the determination of the *evolute*" (pp. 168–79). In his preface he goes even further, praising "such brilliant investigations as those in which, by purely geometrical means, the author arrives at what amounts to the complete determination of the evolute of any conic" (p. vii). In a similar vein, ver Eecke states: "Moreover, most of the propositions of this book approach in a striking way modern theories regarding normals, subnormals and radii of curvature, and one finds already the germ of the theory of evolutes" ("La plupart des propositions de ce livre se rapprochert d'ailleurs d'une manière frappante des théories modernes sur les normales, les sous-normales et les rayons de courbure, et l'on y trouve déjà le germe de la théorié des développées," p. xix).

5. Boyer, *History,* p. 414. Boyer repeats another common myth regarding Huygens's work on evolutes, namely that the Pascal challenge problems motivated Huygens to apply the cycloid to his clock, thereby leading to his analysis of isochronism, a year before it actually occurred (p. 410).

6. Apollonius was also concerned with finding lines of greatest length from a point to an ellipse. I have eliminated references to those cases in order to simplify the explanation. The only modern edition of Book 5 is ver Eecke's French translation based on a Latin translation by Edmond Halley. The most important propositions are numbered 51 to 54 (pp. 418–37 of ver Eecke's edition).

7. See the quotations in note 4; also Boyer's statement, "Neil's parabola was not entirely unknown in antiquity, for Apollonius had recognized its equivalent as the evolute of the parabola" ("Early Rectifications," p. 38).

8. Pappus of Alexandria, *Collectionis quae supersunt,* Vol. 1, Book 4, pp. 271–3, and Vol. 2, Book 7, p. 677. There has been argument among historians regarding the exact interpretation of Pappus' objection; for a summary of the literature on the debate, see ver Eecke's translation, Vol. 1, pp. 208–9, n. 4.

9. The ties between Gool and Huygens were never very strong, particularly compared with those between van Schooten and Huygens. Gool recovered the manuscript in 1629; Mersenne mentions it in his *Universae geometriae* (Paris, 1644) and van Schooten in his *De organica conicarum sectionum in plano descriptione, tractatus* (Leiden, 1646). Constantijn Huygens gives a brief history of its recovery, including the fact that the translation was still not forthcoming, in a letter to Mersenne in 1646 (*OH* 2:555). Christiaan updates the history in 1654 in a letter to Grégoire de Saint-Vincent (*OH* 1:265). There is no evidence that Gool ever completed a translation or even communicated the contents of the books. Halley's edition, *Apollonii Pergaii Conicorum libri octi et sereni antissinsis de sectione cylindri et coni libri duo* (Oxford, 1710), includes (1) the Greek text of the first four books with facing-page Latin translations, (2) Pappus' list of lemmas implicitly used by Apollonius, and (3) a Latin translation of Gool's Arabic manuscript, Books 5–7.

10. Boulliau and Huygens exchanged a series of letters regarding the Medicean manuscript and the status of Gool's translation (*OH* 2:226, 252, 269, 275). For a history of this second manuscript, which ver Eecke conflates with that

of Gool, see Bortolotti's two articles in which he proves (with additional evidence provided by Agostini) that this manuscript was actually transported to Italy in the sixteenth century but remained untranslated for want of expertise and monetary support. The title of the eventual synopsis gives some hint of its checkered history: *Apolloni Pergaei Conicorum, Lib. V, VI, VII, paraphraste Abalphato Asphananensi nunc primum editi Abrahamis Ecchellensis Maronita in alma Urbe linguar, Orient. Prof. Latinos reddit. (Florentiae MDCLXI). Io. Alphonsi Borelli Praefatio.*

11. Problem 5 of *Alexandri Andersoni scoti exercitationum mathematicarum duas prima* (Paris, 1619) states, "Given a Parabola, and given a point outside it, to find a straight line tangent to the given Parabola, such that, when a straight line is drawn from the given point to the point of tangency, it will be perpendicular to the tangent" ("Data Parabola, datoque extra eam puncto, inuenire rectam datam Parabolam tangentem, ad quam, quum ducetur à dato puncto recta in punctum contactus, erit ea ad tangentem perpendicularis," pp. 24–7). Note that he does not express the problem in terms of minimums. Vincenzio Viviani also "reconstructed" this problem in his *De maximis, et minimis geometrica divinatio in quintum conicorum apollonii pergaei adhuc desideratum* (Florence, 1659). Viviani more properly treats the problem as "From a given point, to a given Parabola not containing it, to draw a MINIMAL straight line" ("A dato puncto, ad datae Parabolae peripheriam, MINIMAM rectam lineam ducere," pp. 23–4). He also uses an auxiliary hyperbola in his proof. Huygens received and criticized Viviani's proof in the spring of 1660 (*OH* 3: 60–1, Apr. 7, 1660).

12. Although Huygens argued that the hyperbola could be eliminated, he considered unanswerable the larger question implicit in Pappus' objection, namely whether a problem that has a conic section as a *given* entity is planar or solid. Huygens interprets Pappus to mean it is planar – hence, the objection to introducing a solid (the hyperbola) into the solution. For alternative interpretations see ver Eecke's summary cited in note 8.

13. Huygens briefly describes the meeting with Abraham van Berckel in a letter to van Schooten (*OH* 1:242–3, Sept. 20, 1653). On f. 122v of *Hug.* 12 (dated Sept. 17) Huygens records his proof; f. 123r–v (both dated Sept. 1) contain derivations of the normals to the cissoid and conchoid; ff. 124r and 125r (Sept. 25) hold his derivation of the inflection point of the conchoid; see *OH* 12:76–86 for all the material. Huygens's editors report that inserted in the notebook (f. 123b) is a derivation in a different hand of the tangent to the conchoid. Since Huygens records van Berckel's name on the sheet, it can be assumed that van Berckel presented Huygens with his own proof of a problem that arose in their discussion. Huygens did not send van Schooten his own proof regarding the parabola, since he considered it easy to construct from methods presented by Descartes and van Schooten for the solution of cubic equations. Later he did send it to Carcavy and Mylon (*OH* 1:427–9). A month later van Schooten received Huygens's derivation of the conchoid's inflection point (*OH* 1:243–6, Oct. 23, 1653). Huygens published the latter in *Illustrium quorundam problematum constructiones,* which appeared as an appendix to *De Circuli Magnitudine Inventa;* see *OH* 12:211–15 for reprint

with facing-page French translation. Van Schooten announced Huygens's result regarding the parabola in his second edition of the *Geometria,* 1659; his comments are reproduced in *OH* 14:421-2.

14. Both works are reproduced in *OH* with facing-page French translations: *Theoremata* in 11:281-337; *Circuli* in 12:113-215.

15. In Apollonius' geometric approach, the normals must cross the axis so that he can manipulate the line segment that results.

16. "Toutefois je puis dire avec veritè que je ne m'y suis point attachè, croiant avoir desia plus fait qu'on n'avoit requis, … de sorte que je marquay seulement les proprietez qui se presenterent dans la suite de mon raisonnement, sans m'ecarter a poursuivre d'autres dans l'incertitude de rien rencontrer qui me paiast de ma peine" (*OH* 10:216).

17. The three solutions, all published in the June 1691 issue of *Acta Eruditorum,* were in response to a question raised by Jakob Bernoulli in the May 1690 issue. Both Leibniz and Johann Bernoulli discovered that the construction of the catenary could be reduced to the quadrature of the hyperbola. This result, more than the trial problems Leibniz solved for him, convinced Huygens of the value of the new technique.

18. "The lines were not being sought as normals to the tangent at its point of contact, but rather as the smallest possible or largest possible lines that one can draw from a given point to a conic" ("Die gesuchten Linien werden nicht als senkrecht auf Tangenten in deren Beruhrüngspunkten bestimmt, sondern als die möglichst kleinen oder möglichst grossen Linien, die man von einem gegebenen Punkte an einen Kegelschnitt ziehen kann," p. 293).

19. In quoting Eutocius and Pappus on the *diorismos,* Heath specifically cites the solution of Book 5 as a good example (p. lxx of his synopsis). He also discusses the *diorismos* in his *Greek Mathematics* 1:319, 371.

20. "Si sit curva quaedam *AB,* et alia item curva *CD* illa exterior, hoc est cujus cavitas respiciat convexitatem curvae *AB*" (*OH* 14:398).

21. Huygens uses a variety of phrases such as "versus eandem partem cava" and "linea in unam partem inflexa"; for examples, see *OH* 14:398-403 and 18: 189.

22. In his approach Huygens notes the geometrically obvious fact that the subtangent must be at its maximal value at the inflection point. He sets up an algebraic statement of this fact, then applies Fermat's method of maximum and minimum. Huygens returned to the problem of the inflection point in the hope of finding a proof simpler than both his 1656 proof and one by van Heuraet that he admired, which was published in van Schooten's second edition of the *Geometria,* 1659. For the later derivation, see *OH* 12:232; for editorial comment and further reference, see *OH* 12:110-12.

23. "Now by how much y[e] nigher y[e] points *d* & *f* are to one another, soe much y[e] lesse difference there will bee twixt y[e] crookednesse of y[e] pte of y[e] line *de*, & a circle described by y[e] radius *df* or *ef*. And should y[e] line *df* be understood to move until it bee coincident w[th] *ef*, taking *f* for y[e] point where they ceased to intersect at their coincidence, y[e] circle described by y[e] radius *ef*, & y[e] crooked line at y[e] point *e*, would bee alike crooked" (*Mathematical Papers* 1:253). Newton's earliest derivations of curvature, beginning with the parabola in 1664, appear in *Mathematical Papers* 1:248-71.

24. *Mathematical Papers* 1:265.
25. *Mathematical Papers* 1:272–97, for the draft; 1:400–48, for the 1666 treatise.
26. *Mathematical Papers* 1:263.
27. *Mathematical Papers* 1:432–41.
28. *Mathematical Papers* 2:206–47, for *De Analysi per Aequationes Numero Terminorum Infinitas;* 3:32–353, for *De Methodis Serierum et Fluxionum.*
29. From a letter to Vernon written by Collins and quoted by Whiteside in his introductory notes to *De Methodis* (*Mathematical Papers* 3:23).
30. *OH* 7:328.
31. *OH* 7:332.
32. See note 17. For Huygens's response, including early drafts and letters, see *OH* 9:496–515.
33. *OH* 9:496–9 and 10:95–8.
34. "Item definitur radius curvitatis in vertice *V*, hoc est semidiameter circuli maximi qui per verticem hunc descriptus totus intra curvam cadat" (*OH* 10:96).
35. "Je n'avois pas songé à la courbe, qui par son evolution peut produire la chainette. Cependant je voy qu'il est bon d'y songer dans les recontres. Je ne scay, Monsieur, si vous avés remarqué un petit discours de Angulo contactus et Osculi, que j'avois mis dans les Actes de Leipzig mois de Juin 1686. Où je considere, que la direction de la courbe se doit exprimer par la droite qui la touche, parce que la droite a partout la même direction: Et la droite qui touche ne fait avec la courbe qu'un angle de contact, qui est moindre que tout angle de droite à droite. Mais la courbure ou flexion de la courbe en chaque point se doit exprimer par le cercle qui l'y touche le plus exactement, ou qui la baise, car le cercle a par tout la même courbure; et le cercle qui baise ne fait avec la courbe qu'angulum Osculi, comme je l'appelle, qui est moindre que tout angle de contact de cercle à cercle. Et ce cercle sera la mesure de la courbure. Ce qui s'accorde avec ce que vous dites, Monsieur, du rayon de la curvité. C'est pourquoy on fait bien de consider cecy en examinant les courbes. Et les centres des cercles mesurans la courbure tombent dans votre generatrice par evolution. Il seroit peut-estre bon de continuer la progression et d'examiner quelle courbe seroit la plus propre à estre la mesure de l'osculation du second degré" (*OH* 10:156). For the article, see Leibniz, *Mathematische Schriften* 7:326–9.
36. "Vous me parlez à propos de la courbure de la chainette, de vostre *discours de angulo Contactus et Osculi.* Vous pouvez bien croire qu'en ce lisant je ne trouvay pas cette considération nouvelle, parce que ces fortes de contact entrent naturellement dans mes Evolutions des Lignes courbes.
   "Je me souviens aussi que longtemps devant que de publier ce Traité j'avois communiquè à van Schoten quelque remarque là dessus, scavoir de la circonference, qui coupant une parabole, semble la toucher au mesme point, c'est à dire que dans la parabole comme aussi dans les autres sections coniques il n'y a que le point du sommet où une circonference la puisse baiser; cela arrive encore en plusieurs cas d'autres lignes courbes, quoy qu'il semble que vous n'en avez rien dit" (*OH* 10:183–4). The original letter to van Schooten is found in *OH* 1:305.
37. "Je crois bien que Vous avés vu que le cercle qui se decrit du point de la courbe evolue, et dont le rayon est la moindre droite qu'on peut mener de ce

point à la courbe decrite; mais peut-estre n'aviés vous pas songé d'abord à le considerer comme la mesure de la courbure, et moy lorsque j'avois consideré le plus grand cercle qui touche la courbe interieurement comme la mesure de la courbure ou de l'angle de contact, je ne m'etois pas avisé de songer aux evolutions" (*OH* 10:227).

38. For his confession, see *OH* 10:568.

39. Johann Bernoulli would later make explicit the idea latent in Huygens's approach, when he derived the radius of the osculating circle by means of the same proportion drawn from similar triangles that Huygens used (*Opera Omnia* 3:437). For an analysis of this and other treatments of the radius of curvature during the 1690s, see Bos, "Differentials," pp. 36–42.

40. I have not had a chance to inspect Leibniz's marginalia to his copy of the *Horologium Oscillatorium,* now in the Landesbibliothek Hannover.

41. The appropriate section of *De Circuli Magnitudine Inventa* appears in *OH* 12:165.

42. Yes, dear mathematician, it is technically possible to create a general analysis of curvature from parabolas rather than circles. Given a curve, define its curvature at a point p to be the reciprocal of the *latus rectum* of the "osculating parabola" that has as its axis the normal at p. Because the radius of the osculating circle is half the *latus rectum* of the osculating parabola, the paraboloidal curvature would be twice the circular curvature.

43. "A curve is drawn that touches *AB* in *A*, *BG* in *D*, *GP* in *L*, which I posit in place of the catenary, and which I replace with the circumference of a circle, or even a parabola, and I now investigate the diameter of that circumference" ("Ducta est enim curva quae tangit rectam *AB* in *A*, *BG* in *D*, *GP* in *L*, quam pro curva catenae hic habeo, et quam pro circumferentia circuli, aut etiam parabola, reputo, cujus circumferentiae diametrum hic porro investigo"; *OH* 9:504).

44. For his initial discussion, see Leibniz, *Mathematische Schriften* 7:331–7. In fact, only three-point contact is necessary, as Jakob Bernoulli tried to explain to a stubborn Leibniz. Their exchange took place in the *Acta Eruditorum* in 1692 and is reprinted in Jakob Bernoulli, *Opera* 1:473–81, 543–8.

45. *OH* 10:304–10, 312–15, 325–35; see particularly pp. 333–5 for his proof.

7. RECTIFICATION

1. In other words, today we can say that the square with side-length $\sqrt{\pi}$ has an area equal to the area of a circle of unit radius, but the side of that square cannot be constructed by ruler and compass because pi is transcendental; that is, pi cannot be achieved through a finite series of algebraic manipulations beginning with a rational number. Thus, there is no solution within seventeenth-century standards.

　　Descartes's *Géométrie* was one of three appendixes to his *Discours de la methode* (Leiden, 1637) and as such was meant to show how to apply the method to a specific science; reprinted in Vol. 6 (1902) of *Oeuvres de Descartes.* Van Schooten's *Geometria* (Leiden) first appeared in 1649; an expanded version was published in 1659.

2. Archimedes, *Quadratura parabolae,* in *Opera Omnia* 2 (1913):261–315.

3. In these cases, such as Dinostratus' rectification of the circle using the tractrix and Archimedes' using the spiral, the defects of constructability lay with the curves being used to achieve the rectification, because they were mechanical curves dependent on the circular arc for their actual construction, and thus they were not independent of the solution being sought. See Boyer, "Early Rectifications."

4. Olscamp translation, p. 206.

5. The letter, dated Sept. 12, 1638, was published in *Lettres de Descartes,* ed. Claude Clerselier, 3 vols. (Paris, 1657, 1659, 1667), 1:347–54, Letter 74; reprinted in *Oeuvres de Descartes* 2 (1898):352–62, Letter 142 (passage on spiral, p. 360).

6. The debate began with Hobbes's *Elementorum Philosophiae: Sectio Prima, De Corpore* (1655). Wallis's reply, *Elenchus Geometriae Hobbianae* (1655), prompted Hobbes to change some of his proofs for the English translation (*Elements of Philosophy: The First Section, Concerning Body,* 1656) and to add a lengthy addendum under the title *Six Lessons to the Professors of Mathematics of the Institution of Sir. Henry Savile in the University of Oxford;* reprinted in *The English Works of Thomas Hobbes, Elements* in Vol. 1, *Six Lessons* in Vol. 7, pp. 181–356. To which Wallis replied, etc. For a discussion of the quarrel, see Scott, pp. 166–72.

7. Grégoire's attempted quadrature appeared in Book 10 of his *Opus Geometricum* (Antwerp, 1647). For a synopsis of the work see Hofmann, "Das Opus Geometricum," Huygens's criticism, E'ΞE'TAΣIΣ *Cyclometriae Cl. Viri Gregorii a S. Vicentio* (1647), is reprinted in *OH* 11:314–37.

8. *OH* 1: passim, particularly pp. 495–502 for a letter to Francis Xavier Ainscom. For a history of the argument by Huygens's editors, Ainscom's initial argument, and the published version of Huygens's reply, see *OH* 12:239–77.

9. *Dimensio Circuli,* in *Opera Omnia* 1:231–43.

10. Boyer, "Early Rectifications," pp. 37–8. A similar view is expressed by Hofmann, "Rektifikationen," pp. 283–4.

11. Huygens received a copy of one of Hobbes's attempts to square the circle from William Brereton, who pointedly mentions Huygens's criticism of Grégoire de Saint-Vincent in his cover letter (*OH* 1:333–4, June 23, 1655). (His editors err with regard to which edition he received; it was the original Latin, not the English second edition.) Wallis sent Huygens another copy, stating in his cover letter a few disparaging remarks about Hobbes and, in particular, noting Hobbes's erroneous rectification of the parabola (*OH* 1:335–8, July 1, 1655). Coincidentally, because Huygens was abroad in the fall of 1655, he received both sides of the argument in the same packet of letters. Only the reply to Wallis exists (*OH* 1:392, undated but referred to by Wallis as Mar. 15, 1656). Huygens's aloofness in the quarrel is exemplified by an I-told-you-so gibe made two years later when Wallis was still combating Hobbes (*OH* 2:211), to which Wallis gave a my-honor-was-at-stake rejoinder and reminded Huygens of his own public reply to Ainscom (*OH* 2:296). Huygens obviously was on Wallis's side, telling Moray in 1662, upon receipt of the latest broadside, that Wallis had "pleasingly mocked his man" (*OH* 4:149). It was at this

time that Huygens joined the fray, because Moray cajoled him into assessing Hobbes's latest effort to square the circle as well as to duplicate the cube (*OH* 4:176–7, 200–5, 280–3).

12. There is a possible link to Hobbes's mathematics that I have not researched. Another aspect of the debate between Hobbes and Wallis focused on the "angle of contact" made between a circle and a tangent to it, Hobbes arguing that the angle of contact could be measured and thus used to determine the "quantity of crookedness" (*Six Lessons* in *English Works* 7:260). Could Newton have interested himself in the question of curvature because of this debate?

13. Neil's work is mentioned in the Boyer and Hofmann articles on rectification and more thoroughly described by Christensen.

14. Of course, Wren's solution was not a geometric rectification of a *geometric* curve, as Neil's was, but that objection seems to have been overridden by the fact that it was a direct result for an extraordinary curve.

15. Reprinted in Wallis, *Opera Mathematica* 1:533–9 (Wren's proof) and 550–3 (Neil's parabola).

16. *OH* 2:416–17, June 9, 1659.

17. *OH* 14:234–6, Oct. 27, 1657.

18. "Amo autem ea problemata in quibus inventio praecipua calculus vero facilis. Circa parabolam ante paucos dies duobus novis, ut mihi quidem videntur ac praeclaris inventis potitus sum, quibus conscribendis summo studio nunc incumbo" (*OH* 2:80, Nov. 2, 1657).

19. His letter to Sluse begins: "I told you before that I discovered two new things concerning the parabola; the one I divulge to you, the other I withhold still for certain reasons. I have found by what means a circle equal to the surface of a given parabolic conoid is described" ("Significavi tibi antehac duo nova circa parabolam me deprehendisse; horum alterum tibi edam, alterum certa de causa adhuc reticebo. Inveni itaque quomodo dati parabolici conoidis superficiei circulus aequalis describatur"; *OH* 2:104, Dec. 20, 1657). An almost verbatim copy of Huygens's letter to Sluse is inserted into a longer letter to van Schooten (2:110–13, Dec. 28, 1657).

20. A collection of neat, geometric proofs for the two results (still another small treatise!) is reproduced in *OH* 14:237–70.

21. Only the algebraic expression survives. Huygens annotates it as having come from van Schooten, to whom van Heuraet sent it on Jan. 18 (*OH* 2:131). On Feb. 3 Huygens confirms the expression with his own derivation (*OH* 14:314).

22. "Quod autem alterum tuum inventum concernit, de quo mihi non nisi subobscurè quaedam indicaveras, penitus fermè cum Heuratium adibam jam tunc exciderant mihi, ita ut inde scire non potuerit in quo illud ipsum propriè consisteret, minusque an ejus methodus cum eodem hoc tuo invento responderit. Quocirca petijt ut quale illud tandem sit data occasione ei indicare haud gravatè duceres, quo desiderio tuo satisfacere posset" (*OH* 2:129–30, Feb. 4, 1658). Huygens's editors append the expression cited in note 21 to this letter despite an obvious dating problem.

23. "En soo uEd. myn humeur slechs gekent had 't waer niet nodich geweest soo veel moeyten tegen my aen te wenden die uEd. t'vermaecken en d'eer van geseijde inventie schoon deselve voor lanch van mij gevonden mochten syn geensins sal soecken te benemen" (*OH* 2:138–9, Feb. 24, 1658).

24. Writing to Boulliau, he had earlier declared, "They seem to me so difficult for the most part that I strongly doubt if even he who has proposed them can solve them all, and I would well wish that he had assured us of that in the announcement. Otherwise, it is very easy to invent some impossible problems" ("Ils me semblent si difficiles pour la pluspart que je doubte fort si celuy mesme qui les a proposez les pourroit tous resoudre, et voudrois bien qu'il nous en eust assurè dans ce mesme imprimè"; *OH* 2:200, July 25, 1658). When subsequently he was able to prove another of the problems, he retracted his statement (*OH* 2:220, Sept. 19, 1658).

25. Both his letter to Boulliau and his summary of one to Carcavy are dated Jan. 16, 1659 (*OH* 2:313-17). Mylon replied with the news regarding Auzout (2: 332-5, Jan. 31, 1659), and Carcavy specified the date of Auzout's discovery (*OH* 2:345, Feb. 7, 1659).

26. Huygens's letter to van Schooten includes a complete mathematical statement of the result (*OH* 2:343-5, Feb. 7, 1659). Van Schooten replied within a week (2:352-4, Feb. 13, 1659).

27. Huygens first communicated his "theorem of the parabola" to Sluse on Jan. 14, 1659; see his summary in *OH* 2:312-13. His attempt to subpoena Sluse as witness to his priority came much later (2:417-18, undated copy but probably sent in May). Sluse's reply that he had not seen the theorem on conoid surfaces is dated June 13 (2:422-3). After Huygens repeated his result (2:435-6, undated), Sluse apologized about the mixup, stating that he thought Huygens was referring to some formal announcement (2:436-8, July 15, 1659).

28. *OH* 2:329-31, Jan. 31, 1659.

29. "Is cum rescivisset me Conoidis Parabolici superficiem dimensum esse rectamque lineam parabolae invenisse aequalem supposita hyperbolae quadratura (de quibus antea tibi scripsi) non tantum utrumque horum suo marte invenit, sed et rectas alijs curvis absolutè aequales ostendit, ex earum genere quae in Geometriam recipimus" (*OH* 2:416-17, June 9, 1659). He goes on to cite the specific case of the semicubical parabola. Note that Huygens had only hinted at the conoids result in his previous letter.

30. This is the same letter in which Huygens is trying to stir Sluse's memory about his earlier result (*OH* 2:435-6).

31. "Cum vero in his simus, etiam de nobis dicere liceat, quid ad promovendum tam eximium inventum contulerimus: siquidem & Heuratio ut eo perveniret occasionem praebuimus, & dimensionem curvae parabolicae ex hyperbolae data quadratura, quae Heuratiani inventi pars est, ante ipsum atque omnium primi reperimus. Etenim sub finem anni 1657 in haec duo simul incidimus, curvae parabolicae quam dixi dimensionem, & superficiei conoidis parabolici in circulum reductionem. Cumque Schotenio, aliisque item amicorum, per literas indicassemus, duo quaedam non vulgaria circa parabolam inventa nobis sese obtulisse, eorumque alterum esse conoidicae superficiei extensionem in circulum, ille literas eas cum Heuratio, quo tum familiariter utebatur, communicavit. Huic vero, acutissimi ingenii viro, non difficile fuit intelligere, conoidis istius superficiei affinem esse dimensionem ipsius curvae parabolicae. Qua utraque inventa, ulterius inde investigans, in alias istas curvas paraboloides incidit, quibus rectae aequales absolute inveniuntur" (*OH* 18:211). He goes on to cite a letter from Sluse praising his own accomplishments.

32. Van Maanen accepts Huygens's history and offers a possible reconstruction of van Heuraet's procedure in "van Heuraet."

33. "Huygens nous assure, et il n'y a pas lieu d'en douter..." (*OH* 14:189). I should stress that my argument that van Heuraet's work was completely independent is a minority opinion. In contrast, Hofmann in "Rektifikationen," p. 286, and Struik in his entry on van Heuraet for the *Dictionary of Scientific Biography* accept Huygens's claim of influence, although both erroneously report some of their documentation.

34. Van Schooten died in 1660. Van Heuraet's fate is unclear, although he also appears to have died in 1660; see van Maanen, "van Heuraet."

35. For the text with facing-page English translation, see Grootendorst and van Maanen.

36. In his letter informing Huygens of van Heuraet's work, van Schooten claims more for van Heuraet than is directly shown in the published method. In particular, the printed version does not contain any reference to higher-dimensional curves such as the paraboloid that caused the dispute between Huygens and van Heuraet. Nor are centers of gravity treated, although van Schooten states that van Heuraet mentioned applications involving them. He goes on to state that with this new discovery van Heuraet is to be ranked with the world's great mathematicians, Huygens included (*OH* 2:352–4, Feb. 13, 1659).

37. If $ay^2 = x^3$, then XT/XS (i.e., dy/dx) equals $3\sqrt{x}/2\sqrt{a}$ and CQ/CM, or $ds/dx = \sqrt{1 + (9x/4a)}$. Van Heuraet factors out $9/a^2$, so that

$$\frac{CQ}{CM} = \frac{ds}{dx} = \frac{3}{a}\sqrt{\frac{a^2}{9} + \frac{xa}{4}} \; ;$$

that is, he picks the arbitrary constant W equal to $a/3$, and since $MI/W = CQ/CM$, $z = MI = \sqrt{(a^2/9) + (xa/4)}$.

38. When $y = x^2/a$, XT/XS or $dy/dx = 2x/a$, and CQ/CM or $ds/dx = \sqrt{1 + (4x^2/a^2)}$. Again, factoring and setting $W = a$, van Heuraet has $z = MI = \sqrt{4x^2 + a^2}$.

39. See *OH* 14:451–7 (Aug. 1661) for his "fundamental rule for finding logarithms" from the hyperbola; *OH* 14:474–80 (July 1662) for his determination of the arc length of the parabola. On Grégoire de Saint-Vincent's work, see Naux.

40. *OH* 18:219–21.

41. Huygens received Wallis's treatises on the cycloid and cissoid on Mar. 20, 1660 (*OH* 3:58). Anxious to see van Heuraet's work, Huygens asked van Schooten to send him a prepublication copy of the *Geometria,* or at least the pages containing van Heuraet's work, to which van Schooten agreed (*OH* 2:412–13, 415, June 1659).

42. Additional proof that rectification was still on Huygens's mind in the fall of 1659 is furnished by a letter that he sent to Grégoire de Saint-Vincent on Oct. 30, which contains exact mathematical statements of his two discoveries regarding the parabola (*OH* 2:500–2). Note that he was very freely communicating with his "rival," which says something about the need of historians to keep priority squabbles in perspective.

43. *OH* 17:122. He claims that his method is shorter than van Heuraet's. In fact, there are more differences than that, which leads me to suspect that he had not yet received van Heuraet's theorem from van Schooten (see note 41).

44. "Proinde ad rectas lineas reduci possint" (*OH* 17:146). This passage is indirect evidence of an early existence of his general theorem on evolutes. Although only the version printed in the *Horologium Oscillatorium* survives, he refers in this passage to a figure whose line segments are labeled with the same lettering as the printed theorem. Given his penchant for relettering figures in each new proof, it seems highly likely that he had the draft of the theorem lying before him.

45. *OH* 18:237–41.

46. "Data linea curva, invenire aliam cujus evolutione illa decribatur; & ostendere quod ex unaquaque curva geometrica, alia curva itidem geometrica existat, cui recta linea aequalis dari possit" (*OH* 18:225).

47. *OH* 18:203–5.

48. "Excellent om dat de eerste kromme is en misschien d'enighste die kan rect gemaeckt werden" (*OH* 2:315, Jan. 16, 1658 [*sic*]). The quotation comes from a draft of a letter to Carcavy, Pascal's intermediary in the Dettonville problems.

49. The letter is quoted more fully in Chapter 5, note 26.

50. "Circa lo Oriuolo regolato dal Pendolo, certo è che l'Invenzione è bella, ma non si deve defraudare della gloria douutali al nostro Signore per sempre ammirabile Galileo, che gia nel mille seicento trentasei, si io non erro, propose questa si utile invenzione alli Signori Stati d'Olanda et io ne ho ritrovato, benche in parte diverso circa la constituzione delle ruote, un modello fatto gia dal medesimo Signore Galileo" (*OH* 3:462, Mar. 31, 1659). Boulliau copied out this passage and sent it to Huygens (*OH* 2:404, May 9, 1659).

51. "J'ay respondu sur cela a Son Altesse Serenissime que je scauois que vous tiendriez a honneur, & que vous croirez meriter de la gloire, si vous estez tombé dans les mesmes pensees que Galilei a eues; & que vous estiez si homme d'honneur & si sincere que vous ne defrobberez jamais la reputation d'autruy pour vous lattribuer, vous auez de l'esprit au dela de l'ordinaire fertile en de tres belles inuentions, & ainsi pour vous satisfaire, & pour vous acquerir de la renommee vous n'auez pas besoin des inuentions d'autruy" (*OH* 2:403, May 9, 1659). Boulliau's letter to Leopold is dated May 2 (*OH* 3:462–3). Boulliau actually began the dispute by sending one of his own copies of the *Horologium* on to Florence (*OH* 2:252–3, Oct. 18, 1658).

52. "De sorte que la pensée semble avoir estè assez commune, mais on ne peut nier que mon modelle n'ait succedè le premier" (*OH* 2:405, May 14, 1659).

53. "Certes je me croirois indigne de vie, mais puis que pourtant la negative est difficile a prouuer, je ne voy pas quelle autre apologie je pourrois faire a Son Altesse Serenissime que de luy protester avec toute sinceritè que ny moy ny personne jamais en ce pais, à ce que j'ay pu apprendre, n'a ouy parler de cette invention devant que je la produisasse" (*OH* 2:406). Boulliau sent excerpts of this letter to Leopold (*OH* 3:466, June 13, 1659).

54. *OH* 2:430–4.

55. "Quando io le accennai che l'invenzione d'adattare il Pendolo era stata trovata molto tempo fà ancora dal Nostro Signore Galileo, non intesi di dire che il Signor Christiano Hugenio non la potessi haver' anch' egli ritrovata da se medesimo, sapendo molto bene l'Eminenza del suo Ingegno e dottrina mediante le quali cose bene puo concludersi che questo gran Virtuoso possa haver' ritrovato questa e ritrovare ancora cose maggiori" (*OH* 3:464, May 22,

1659). This passage was also transcribed by Boulliau and forwarded to Huygens (*OH* 2:431-2, July 4, 1659).

56. Huygens tells Boulliau that he was surprised to receive praise rather than the rejection of his defense that he had expected (*OH* 2:441, July 24, 1659).

57. Leopold's copy was sent Aug. 19, 1659 (*OH* 2:453). On Nov. 20, 1659, Huygens tells Boulliau that he is still waiting for a reply, although he knows from Heinsius that Leopold has received and praised the book (*OH* 2:509-10). His complaints continue into the new year (*OH* 3:12-13, Jan. 22, 1660).

58. Because Huygens had not sent a personal cover letter with the *Systema Saturnium,* the prince would not respond personally, even though the book was dedicated to him; see Huygens's report to Chapelain (*OH* 3:119, Sept. 2, 1660). Accustomed to a different court's manners, Huygens believed that a personal letter "would make it seem that I requested something beyond just his approbation" (*OH* 3:31, an annotation to a letter from Boulliau dated Feb. 27, 1660). He apparently waited a while longer for a response, having heard that Leopold was sending his approval indirectly (via Dati to Heinsius), but when nothing materialized, he finally wrote an appropriately obsequious letter directly to Leopold – sent via Heinsius; see *OH* 3:32, 502-13 for the background, *OH* 3:109-10 for the letter (Aug. 13, 1660).

59. Fabri and Divini were his main opponents (*OH* 3:passim, particularly pp. 129, 147-68, 195-8). For a history of the debate, see van Helden, "Divini versus Huygens," and "The Accademia del Cimento."

60. *OH* 3:467-97, letters dated Aug. 21 and Oct. 9, 1659.

61. On Dec. 19, 1659, Boulliau reports to Leopold that he has communicated with Galileo's friend, but his account relates only generalities regarding how highly the man still values that friendship (*OH* 2:531). In passing, Boulliau mentions that he will forward Viviani's history to Huygens, but there is no evidence that he did. After being shelved by Boulliau, Viviani's account languished for more than a century, its very existence in doubt; for more on the history of this work itself, see *OH* 3:470-2, n. 1, and both Dobson and Edwardes. Leopold did revive the priority claim in 1673, but with little vigor (*OH* 7:279-86).

62. Boulliau sent a copy of Galileo's design on Jan. 9, 1660, and a drawing of the extant clock on Jan. 23 (*OH* 3:8, 14). Huygens's assessment was nominally based on the first drawing, although it obviously was a foregone conclusion (*OH* 3:12-13, Jan. 22, 1660). With regard to the second drawing, Huygens charged that the actual clock was still too poorly designed to function accurately, although it had been modified to alleviate some of the difficulties of Galileo's design (*OH* 3:21, Feb. 12, 1660).

63. *OH* 3:195-8, Nov. 28, 1660.

64. *OH* 18:213-19.

65. *OH* 2:passim. Once again, Mersenne's lingering influence is apparent, for Huygens refers both van Schooten (*OH* 2:111) and Sluse (*OH* 2:115) to the *Tractatus Mechanicus,* in which Mersenne lists without proof propositions concerning parabolas of higher order.

66. *OH* 2:79-80.

67. *OH* 3:113, Aug. 20, 1660.

68. "Alterum est me has omnes curuas, ipsumque adeo locum linearem integrum nihil pene facere, prae inuento hoc tuo, quo superficiej in conoide parabolico rationem ad circulum suae basis demonstrastj. Hanc pro ciculj quadratura pulcherrimam α'παγωγὴν, prefero libenter ijs omnibus quas ex loco lineari nec paucas olim deduxj; et quas tecum sj ita iusseris, data occasione communicabo" (*OH* 2:107, Dec. 24, 1657). In his reply Huygens did, for a change, politely inquire about his correspondent's methods and, at the same time, hinted that he had a similar reduction concerning the quadrature of the hyperbola, i.e., the infamous "other discovery" that he had not yet revealed in full (*OH* 2:116, Jan. 1658).

69. *OH* 18:211–13.

70. As might be expected, the names attached to these expressions have many rivals for the honor; e.g., both Gregory and Newton also knew "Leibniz's" infinite sum.

71. Gregory claimed to have proved the impossibility of squaring the circle in his *Vera Circuli et Hyperbolae Quadratura* (Padua, 1667). He was right, but Huygens, ever the examiner in these matters, found an error in his proof.

72. "Ego vero quomodo quadratus fiat circulus, nec didici, nec praescribo; sed hoc urgeo, ut quem ille modum se invenisse contendit, eum reapse utilem & efficacem esse demonstret" (*OH* 1:501, Oct. 2, 1656).

73. *OH* 14:234–70. Huygens's rapid jump from measuring the arc length of the parabola to measuring the area of the paraboloid's surface depended upon knowledge gleaned eleven years earlier when, after reading Mersenne's *Tractatus Mechanicus,* he realized that the two problems were mutually dependent; see his report to Mersenne (*OH* 1:34, Nov. 1646). Of course, in 1646 he could not solve either problem, but in 1657, after the first had been solved, the second followed automatically.

74. "Parabolici conoidis superficiem ad circulum redigi posse antea tibi significavi. Nunc vero Circuli quadraturam datam esse scito, si sphaeroidis oblongi superficiei circulum aequalem invenire possimus. Quadraturam vero Hyperbolae si superficiei sphaeroidis lati seu compressi circulus aequalis habeatur, vel superficiei conoidis Hyperbolici. Et contra" (*OH* 2:134, Feb. 15, 1658).

75. Sluse raised the topic of the cissoid while discussing the conchoid and other unbounded areas (*OH* 2:144, Mar. 4, 1658). He sent his erroneous proof on Mar. 14 (*OH* 2:150–2). In still another example of his tendency to respond to questions asked by others, Huygens found the area under the cissoid, along with the volume and center of gravity, as soon as the second letter was received (*OH* 14:309–12, Mar. 18). He first stated his results to Sluse and Wallis, then sent the proofs [*OH* 2:164 (April 5) and 178–80 (May 28) to Sluse; 210–14 (Aug. 6) and 329–30 (Jan. 1659) to Wallis].

76. *Opera Mathematica* 1:906 for the cissoid, and 555–9 for the conoids.

77. Huygens's workbook for 1659 contains the beginnings of still another brief treatise, this one on the quadrature of the conoids (*OH* 14:337–46).

78. *OH* 18:209–21.

79. *OH* 2:131.

80. Two separate sheets bear the date Feb. 3, 1658, indicating that the work was probably done in one intense session (*OH* 14:314–36).

81. Bos, "Lemniscate," p. 8.
82. Bos, "Lemniscate," p. 13.
83. See Lectures 8 to 12 of *Geometrical Lectures;* Baron gives a modernized summary, pp. 239–52.
84. See Le Noir; also Boyer, *Analytic Geometry.*
85. Huygens's editors do not indicate that the remark is an addition written in the margin of Huygens's notebook (*Hug.* 10, f. 98r; *OH* 14:395). The derivation of the auxiliary curve is on f. 100v; *OH* 14:396–7.
86. Quoted in Newton, *Mathematical Papers* 7:199.
87. "Je souhaitterois aussi de pouvoir tousjours reduire les quadratures aux dimensions des lignes courbes, ce que je tiens plus simple" (*OH* 10:160, to Huygens, Sept. 21, 1691).
88. Bos, "Differentials," especially Chapters 1 and 2.
89. In 1674, soon after receiving his copy of the *Horologium Oscillatorium* from Huygens, Leibniz states without derivation that the square of the arc length of the circle is equal to the diameter times the arc length of its involute; see *Sämtliche Schriften* 1 (1976):116.
90. Unfortunately for Huygens's sense of priority, Fermat had already rectified the higher-order conics, including the semicubical parabola (making him co-discoverer with Neil and van Heuraet). With the author acknowledged only by the notation "M.P.E.A.S.-Tolosae," Fermat's "De linearum curvarum cum lineis rectis comparatione dissertatio geometrica" appeared soon after Huygens's own discoveries, as an appendix (sep. pag.) to Antoine de Lalouvère's *Veterum Geometria promota in septem de Cycloide libris et in duabus adjectis Appendicibus* (Tolouse, 1660); reprinted in *Oeuvres de Fermat* 1 (1891): 211–54.

8. DIVERSIONS

1. Rather, Huygens refers to a "certain curve"; for the letter to Tacquet, see *OH* 3:1–3, Jan. 1, 1660.
2. *OH* 3:12–13, to Boulliau, Jan. 22, 1660; 26–8, to Carcavy, Feb. 26; 58, to Wallis, Mar. 31.
3. "Je voy plus que jamais par ce dernier escrit de Wallisius les inconvenients et disputes qui en peuuent naistre lors que des inventions de quelque consequence vont de main en main devant que d'estre publiées. Apres que j'auray achevè cela, je vous feray aussi veoir ma methode de mesurer les lignes courbes la quelle je ne croy pas que personne encore s'est imaginée" (*OH* 3:57, Mar. 27, 1660).
4. See, e.g., his continuing revisions of his proof of the isochronism of the cycloid (*OH* 17:138–41 and 18:171–87).
5. *OH* 3:114–15, Aug. 26, 1660.
6. *OH* 3:118–20, Sept. 2, 1660.
7. Friends and influential people, such as the Duke de Roannes (who qualified under both categories), were told, obviously. But so were casual acquaintances, such as a professor of mathematics named Cheaveau, who was also told about the conoid discoveries; see Huygens's travel diary, *OH* 22:536–49.

8. *OH* 3:195–8, Nov. 28, 1660.
9. He records seeing Martinot on Nov. 4 and 11 and on Dec. 1, at which time he told him of the cycloidal clock (*OH* 22:534–9).
10. "Pour Monsieur de Monmor il m'a cent fois demandé ce que je scauois de vous et si je n'auois point eu de vos lettres. On se souuient tousjours de vous en son Assemblée et vous y aués laissé vne odeur de probité qui durera autant qu'elle. . . . Mandés moy surtout si vous donnerés bientost vostre Traitté du Pendule amplifié. . . . Soit sur ce sujet la foit sur d'autres il ne faut demeurer en si beau chemin et vous deués au Public tout ce que vous estes capable de luy donner pour son instruction et pour ses auantages" (*OH* 3:273–4, May 30, 1661).
11. He reported to Prince Maurits of Nassau on Apr. 2, 1661, and was off to Gresham by Apr. 6 (*OH* 22:567–70).
12. See the next two sections of this chapter for more on the Royal Society's decision to test his clocks.
13. *OH* 3:210.
14. "23. disnè en ma chambre. ap.d. s'assemblerent chez moy M. Morre. Mil. Brouncker. S^r. P. Neal. Dr. Wallis, M. Roock, M. Wren. D. Godart. . . . Resolus les cas qu'ils me proposerent touchant les rencontres de deux spheres" (*OH* 22:573, Apr. 23, 1661). Note that the meeting took place in Huygens's room. Later, when the Royal Society launched a formal study of percussion and Huygens became upset that treatises by Wallis and Wren received more attention than his own work, he cited this meeting as evidence of his priority, if not outright influence. To compound matters, Oldenburg, who was not present at the meeting, assumed that the discussion would have occurred during a session at Gresham and was skeptical of Huygens's claim, at one point asking Moray if he remembered any experiments regarding impact. Though apologies and reparations were eventually tendered, the argument significantly cooled Huygens's relationship with the Royal Society.
15. Charles II's formal recognition of the Royal Society occurred on July 15, 1662, and Huygens was made a member during his next visit to England; see Birch, *History* 1:263, June 22, 1660 (o/s). Of course, he was a founding member of the Académie Royale.
16. The other two worthies as far as Oldenburg was concerned were Hevelius and la Chambre (the king's physician) [*OH* 4:367, to Boyle, June 22, 1663 (o/s)]. Also on the list were Viviani, Petit, Carcavy, and Chapelain.
17. "Au Sieur Huggens, Hollandois, grand mathematicien, inventeur de l'horloge de la pendulle . . . 1200" (*OH* 4:405–6 contains the entire list).
18. A continuous stream of letters passed between Huygens and Chapelain regarding the status of the new treatise and Huygens's relationship to the French court (*OH* 5:passim).
19. The whole argument of this paragraph comes from his summary for the now missing letter to his father (*OH* 5:265, Mar. 12, 1665).
20. Lodewijk's letter is missing; Christiaan's reply is dated Dec. 18, 1660 (*OH* 3:209–10).
21. Apparently the clock sent with Lodewijk was a simple modification of the land-based design, because Huygens began making plans for a true marine

version only after his own return home a year later; see his letter to Moray (*OH* 4:65, Feb. 24, 1662) and Moray's skeptical response (*OH* 4:93–4, Mar. 16).

22. The dispute raged over many years (*OH* 4:278ff. and 5:passim) beginning with Huygens's initial complaint to Lodewijk (*OH* 4:278, Dec. 14, 1662), through his appeal to Moray (*OH* 5:6–7, Jan. 9, 1664), the revelation that he could not hold the patent (*OH* 5:116–17, Sept. 23, 1664), to the actual patent grant (*OH* 17:176–7, Mar. 13, 1665).

23. Bruce sends his report Jan. 12, 1663, with a follow-up on Jan. 26 (*OH* 4:290–1, 301). Moray also sends an account (4:296, Jan. 19, 1663).

24. The voyage lasted from Apr. 28 to Sept. 4, 1663, during which time Huygens was again in Paris (*OH* 4:427–8, 431–5, 443–53).

25. *OH* 5:204–6, Jan. 23, 1665.

26. *OH* 5:224–5, Feb. 6, 1665.

27. Moray reports on his continuing investigation in *OH* 5:234, 245, 269–70, 284.

28. *OH* 5:212–15, Jan. 30, 1665; and 17:176, Mar. 13, 1665.

29. Actually, Huygens wrote *Kort onderwijs aengaende het gebruijck der Horologien Tot het vinden der Lenghten van Oost en West* before receiving news of Holmes's second trip and published it soon after Moray's preliminary report of Jan. 1665; reprinted in *OH* 17:199–235, with the summary on pp. 230–5. The English edition was delayed, for this as well as other reasons, until May 10, 1669, when it appeared in the *Philosophical Transactions* No. 47, 4:937–53. Under Chapelain's urging, the French translation that Huygens was supposed to make was withheld (*OH* 5:425). It never did appear, being superseded by the *Horologium Oscillatorium*.

30. The extract from Moray's letter plus a reply from Huygens appeared in French in the *Journal des Savants* No. 8, Feb. 23, 1665, and in an English translation in the *Philosophical Transactions* No. 1, Mar. 6, 1665; see *OH* 5:204–6 for reprints of both.

31. On Dec. 13, 1661 (o/s), Moray cites the proposal as having been made in the session of fifteen days earlier (*OH* 3:427). Birch makes no mention of the subject in his report of the Nov. 27 (o/s) meeting, but does have one line under Nov. 20 referring to Wilkins reading a paper on "a natural standard" (*History* 1:54).

32. "Vous pourrez peut estre, auec iustice, nous reprocher l'impatience qui nous pousse a faire cet experiment deuant que Vostre traitté soit publié. Mais ne nous reprochez point l'impatience puisque, consideré l'enuie que nous auons de voirvos traittez" (*OH* 3:428, Dec. 23, 1661).

33. "Les Experiences que nous auons faites de vostre ligne nous ont reussi a merveilles de sorte que tout le monde en est bien fatissait. neantmoins quant a l'exactitude precise du mouuement du pendule de l'Horologe Je serois aise de scauoir si vous pouuez obseruer qu'il soit tousiours si egal que les changements qui arriuent dans la constitution de l'air n'y apportent nul desordre" (*OH* 4:27, Feb. 3, 1662).

Birch reports that on Jan. 22 (o/s) Brouncker planned the experiments to test the cycloidal pendulum and that Boyle, Petty, Wilkins, and Wren were appointed to the testing committee (*History* 1:70).

34. "Je crois vous auoir dit dans ma precedente que des balles de plomb de differente grandeur ne font pas des vibrations egalles en temps, quelques petites

qu'en soyent les excursions, selon ce que nous auons experimenté, quoy qu'il semble que vous n'y ayiez point trouué de difference ny en differentes matieres, ny en differentes grandeurs" (*OH* 4:35, Feb. 9, 1662).

35. Birch, *History* 1:75, from the meeting of Feb. 5, 1662 (o/s).

36. See Moray's reports concerning the device, which also used a pendulum (*OH* 5:115, 141–2, 156).

37. Huygens's derivations are in *OH* 18:331–3. He told Moray of the rule on Nov. 21, 1664 (*OH* 5:149).

38. Birch, *History* 1:500, 505.

39. *OH* 5:172, Dec. 25, 1664.

40. The pertinent passages from his workbook are reprinted in *OH* 17:183–7.

41. "Cette decouuerte ne m'a pas peu rejouie, estant en mesme temps une belle preuve de la justesse de ces horologes, puis qu'il faut si peu de chose pour les maintenir dans un accord perpetuel" (*OH* 5:247, Feb. 27, 1665). Huygens used similar words when describing the effect to his father (*OH* 5:243–4) and to Sluse (*OH* 5:241).

42. *OH* 5:255–6 and 259–62; both men wrote on the same day, Mar. 6, 1665.

43. Birch, *History* 2:21, from the meeting of Mar. 8, 1665 (o/s). A week later, responding to Pepys's and Moray's latest reports concerning Holmes, Hooke presented a further list of objections, particularly the argument that the vertical component of a ship's motion would always prevent accuracy because, as the ship fell, the bob would not immediately participate in the drop but would temporarily hang in the air, its cord slack and out of control (*History* 2:23–4). Coincidentally, Moray had made the same argument privately to Huygens two years earlier (*OH* 4:318).

44. *Journal des Savants* No. 11, Mar. 16, 1665; reprinted in *OH* 5:244. Huygens tells Moray of his embarrassment, thanking him for not being so hasty (*OH* 5:282).

45. "On a escrit de vos quartiers, par enuie sans doute, que vous mesme auiés trouué des defauts dans vostre Inuention qui vous empeschoient de la publier" (*OH* 5:398–9, July 9, 1665).

46. *OH* 5:440, Aug. 20, 1665. The official offer from Colbert on behalf of Louis XIV was forwarded by Chapelain a week later (*OH* 5:472–3).

47. He overwintered at The Hague and did not actually leave for Paris until Apr. 21, 1666, a week after his thirty-seventh birthday.

48. Mersenne first addressed the problem of finding the "center of agitation" of a triangle suspended at one vertex to Christiaan's father (*OH* 1:23, Oct. 12, 1646). He then repeated it to Christiaan directly as a specific case of the problem of suspended bodies (*OH* 1:45–7, Dec. 8, 1646; and 1:50–6, Jan. 8, 1647). Eighteen years later, when he had solved the problem, Huygens had such pride in his achievement that he returned to the letter to record the date of his solution.

49. *Hug.* 10, f. 89r–v, ca. Dec. 1, 1659; reprinted in *OH* 16:384–91.

50. *Hug.* 10, ff. 91r–101v.

51. *OH* 16:414–33, fall 1661.

52. In response to Moray's news of the impending study, Huygens tells him of his new way to adjust the clock "very precisely" (*OH* 3:438, Dec. 30, 1661).

53. *OH* 16:434–555, with another brief treatise beginning on p. 499 (Oct. 1664).

Huygens tells Moray of this work even before he has completed it (*OH* 5:120, Oct. 10, 1664).

54. After a misunderstanding with Oldenburg, Huygens's seven rules for the motion of colliding bodies were finally published in the *Philosophical Transactions* No. 46, Apr. 12, 1669 (o/s), after having first appeared in the *Journal des Savants,* Mar. 18, 1669; reprinted in *OH* 6:429–33 and 383–5 respectively. Huygens's report on gravity was read to the Académie Royale on Aug. 28, 1669, and the revision, *Discours de la cause de la pesanteur,* was published in 1690; reprinted in *OH* 21:445–87.

55. New clocks went with the Duke of Beaufort to Crete, 1668–9, and with Jean Richer to North America, 1670–1, with negative results in both cases (*OH* 6:218, 378–9, 486, 501–3, and 7:26–7, 54–5, 117). It is a tribute to the accuracy of Huygens's clocks that when, on an expedition to Cayenne in 1672–3, Richer had to adjust his clock even on land, Richer accused the land, not the clock – he conjectured that gravity varies with latitude.

56. *OH* 5:189, Jan. 2, 1665.

57. His illness was severe enough to prompt him to give his papers to Vernon with instructions to deliver them to the Royal Society (!) should he die; see Vernon's report of the "deathbed" scene in *OH* 7:9–13. His sister and cousin cum brother-in-law fetched him home from Paris; see the ensuing letters on the family's concern for his health.

58. On the awkwardness, yet acceptability, of his position, see Belaval.

59. Brouncker had been sending his attempts to Huygens, who found errors in the first two. The third was published in the *Philosophical Transactions* No. 94, May 19, 1673 (o/s), which Oldenburg pointedly noted was before copies of the *Horologium Oscillatorium* had arrived in England (*OH* 7:304–5). Pardies's proof is appended to his *La statique ou la science des forces mouvants* (Paris, 1673). For more on his work see the Ziggelaar articles.

60. *Mathematical Papers* 3:420–31. Whiteside dates the manuscript to 1671, at the same time citing David Gregory's claim that it was done in 1669, antedating the publication of Huygens's results. Did Newton see a clock built to Huygens's design in the late 1660s and reconstruct the mathematics from that? Westfall avoids the problem of explaining Newton's presumption of the cycloid by dating the manuscript to after the receipt of the *Horologium Oscillatorium* (*Never at Rest,* p. 257).

61. *OH* 7:325–32, July 3, 1673.

62. *OH* 18:360–1.

63. For Newton's parallel development of many of Huygens's discoveries, see Herivel, particularly pp. 129–32 and pp. 183–207 for his work on the conical pendulum and circular motion.

64. Newton, *Correspondence* 3:155–6.

65. For a thorough account of Huygens's influence on Leibniz's growth, see Hofmann, *Leibniz in Paris,* esp. Chapter 5.

66. "Vos presens me sont precieux, et je puis dire, que celuy que vous me fistes à Paris de votre excellent ouvrage sur les pendules a esté un des plus grands motifs des progrés que j'aye peutestre faits depuis dans ces sortes de sciences" (*OH* 9:521, Oct. 1690).

67. *Medicina Mentis, sive tentamen genuinae Logicae, in qua disseritur de Methodo detegendi incognitas veritates* (Amsterdam, 1687).
68. Tschirnhaus insisted that he had done the work (on the catacaustic of the sphere) independently (*OH* 8:379–84, Aug. 1682; and 8:463–4, Aug. 30, 1683). Cassini, Mariotte, and la Hire published their report in the *Journal des Savants* of June 8, 1682. Tschirnhaus persisted and published a revision of his work in the *Acta Eruditorum* (*OH* 9:134–52, 499).
69. "Saw that he had taken my Invention of circular pendulum and for falling bodys," Hooke records in his diary (p. 45, May 30, 1673). Hooke lodged a protest with Constantijn Huygens (*OH* 7:395, 390–2, 417–8). Pugliese provides manuscript evidence that Hooke did indeed conceive of a conical pendulum constrained by a semicubical parabola to move on a paraboloid. He is at a loss, however, to explain how Hooke, with no evident mathematics, managed to find the semicubical parabola and opts for Patterson's conjecture that Hooke probably guessed it, given that the curve was popular at the time because of Neil's work. Of course, Hooke's discovery occurred after Huygens's visit to London in 1661, at which time the latter claimed to have communicated this result, as well as his results regarding percussion, to Wren. Wren, who collaborated with Hooke on the pendulum work, denied being told anything by Huygens. Undoubtedly, from Huygens's point of view, this was a convenient lapse of memory that allowed both Wren and Hooke to claim independence in their work on percussion and horology. For more on Hooke's work and the priority dispute, see the articles by A. R. Hall and Patterson.
70. The debate began in 1682 with an article by Catelan in the *Journal des Savants* and reached its peak in 1692 with l'Hospital taking on the primary defense of the new mathematical physics; see *OH* 8:349ff. passim, and Vols. 9 and 10, passim.
71. *OH* 8:115 (Oct. 29, 1678), 125 (Nov. 19), and 189 (July 1679) for de Vaumesle's inquiries; *OH* 18:399–405 for Huygens's work, read to the Académie on Dec. 3, 1678.
72. *OH* 18:489–98.
73. *OH* 7:408–25 and 18:522–6 (1675).
74. Recall his letter to Petit in which he explains that he had originally abandoned the curved plates on the 1657 clock because the pendulum's swing became irregular if the clock was tilted. See his description of the marine clock in the *Horologium Oscillatorium,* which includes a short history of its sea trials (*OH* 18:114–23).
75. Pendulum cylindricum trichordon, 1683 (*OH* 18:527–35).
76. *OH* 19:178.
77. Balancier marin parfait, 1693 (*OH* 18:536–8, 546–61).
78. Libratio isochrona melior praecedente, Mar. 1693 (*OH* 18:562–70). The second variant was the Libra isochronis recursibus, Mar. 1694 (*OH* 18:571–91). For an analysis of this clock as well as a discussion of the others, see Crommelin, "Clocks of Huygens."
79. For a thorough presentation of his work, see both Shapiro articles.
80. "Curva omnium undarum particularium tangens communis erit propagatio undae principalis intra vitrum. Ergo rectae quae hanc curvam tangentem

communem secant ad angulos rectos, erunt radij refracti. Hi autem aliunde dantur. Ergo hi ipsi curvam illam secant ad angulos rectos. Ergo curva oritur ex evolutione curvae alterius quae tangens communis est horum radiorum.

"Sufficit scire quod undae intra vitrum propagantur per istas rectas. Sed cum debeant rectae secare undas ad angulos rectos, mirum videri posset quomodo lineae non ad unum centrum tendentes, undas secare possint semper ad angulos rectos. At hoc jam explicatur per evolutionem" (*OH* 19:422, work on 416–23).

81. *OH* 19:427–31, Aug. 6, 1677.

82. "Ayant montré l'invention de ces lignes courbes qui servent au parfait concours des rayons, il reste à expliquer une chose notable touchant ce que nous avons dit plusieurs fois, que les rayons de lumiere sont des lignes droites, qui coupent les ondes, qui s'en repandent, à angles droits. . . . quelles pourront estre les ondes de lumiere dans ce diaphane, qui soient coupées à angles droits par les rayons convergents? . . . & que deviendront ces ondes apres que lesdits rayons commencent à s'entre couper?" (*OH* 19:534).

83. *OH* 19:535.

84. See, e.g., Johann Bernoulli, "Solutio Curvae Causticae," 1692; Jakob Bernoulli, "Additamentum ad solutionem curvae causticae fratris Jo. B.," 1692, and "Curvae dia-causticae, Natura osculorum uberius explicata," 1693.

85. "Qui Physicam conscribere suscipit sine cognitione matheseos, nae is nugas agit" (Leibniz, *Mathematische Schriften* 3:860).

9. CONCLUSION

1. Bernoulli proposed the problem in the *Acta Eruditorum* of May 1690 (*OH* 10:95–8, 216–18).

2. L'Hospital drew his attention to the question on Mar. 22, 1694, and Huygens replied on June 16 (*OH* 10:585–7, 624–5).

3. "Il me semble que j'avois trouvé que cela estoit autrement, mais vostre authorité fera pour moins que je repete l'examen" (*OH* 10:496–7, Sept. 3, 1693).

4. See van Helden, "Saturn and His Anses" and "'Annulo Cingitur,'" as well as his article in *Studies on Huygens*.

5. Hooke's instrument broke, and the only positive result was his discovery that St. Paul's steeple (from whence he let loose his falling bodies) was 50 feet higher than generally accepted (Birch, *History* 1:461, 467, Aug. 1664). Huygens's only comment was that Riccioli had already gotten the best possible data with that kind of experiment (*OH* 5:101, to Moray, Aug. 8, 1664).

6. *OH* 3:438, Dec. 30, 1661.

7. His attempt to deal with air resistance is documented in *OH* 19:102–57.

8. His summary of a letter to his father states that Hooke "n'entend point la geometrie. se rend ridicule par sa vanterie" (*OH* 5:240, Feb. 19, 1665). To Moray he writes: "Au reste je suis raui de posseder a la fin le dit livre de Monsieur Hook que je ne m'estoit pas imaginè estre un volume de telle importance. Certainement c'est un tres bel ouurage et aussi curieux qu'il ne s'en est imprimè de long temps. Je prends si grand plaisir a le fouilleter qu'a peine

je m'en suis detachè pour vous escrire ces lignes. L'on ne peut pas donner des observations plus exactes en ce genre, ni des figures mieux faites qui assurement luy ont coustè une peine incroiable tant a dessiner comme a faire si bien executer au graveur" (*OH* 5:282, Mar. 27, 1665).

9. For example, *De Vi Centrifuga* and *Dioptrica* were created from the draft treatises for the *Opuscula Postuma,* ed. Burchard de Volder and Bernard Fullenius (Amsterdam, 1703).

10. The Latin *Cosmotheoros* is published with a facing-page French translation in *OH* 21:677–821. The arguments on the universality of mathematics are on pp. 748–51.

11. Dijksterhuis, *Mechanization,* p. 377.

12. "Quicquid supposueris non impossibile sive de gravitate sive de motu aliave re, si inde probes aliquid de magnitudine lineae superficiei vel corporis hoc verum erit. velut Archimedes quadraturam parabolae, posito nisu gravium per parallelas" (*OH* 17:286).

13. "Mais il est vray aussi que je n'avois pas beaucoup meditè alors ces matieres, m'estant tousjours plu d'avantage à chercher l'utilitè de la Geometrie dans les choses de physique et de mechanique" (*OH* 10:308, Aug. 27, 1692).

14. "Vous croiez, à ce qu'il semble, qu'il ne seroit pas extremement difficile d'achever de tout point la Science des Lignes et des Nombres. En quoy je ne suis pas jusqu'icy de vostre avis, ni mesme qu'il seroit à souhaiter qu'il ne restast plus rien à chercher en matiere de Geometrie. Mais cette etude no doit pas nous empescher de travailler à la physique, pour la quelle je crois que nous scavons assez, et plus de geometrie qu'il n'est besoin; mais il faudroit raisonner avec methode sur les experiences, et en amasser de nouvelles, à peu pres suivant le projet de Verulamius" (*OH* 10:190, Nov. 16, 1691). His appeal to Bacon's empiricism might sound gratuitous in light of his own practices, but apparently he did feel the method had value for the scientific community and recommended its approach to the Académie Royale.

15. "Je suis de vostre auis que la geometrie n'est qu'vn ieu d'esprit si on ne l'applique à la phisique et aux inuentions de mecaniques, mais il est rare, qu'on y reussisse et il faut des siecles entiers pour produire vn Hugens" (*OH* 10:393, Feb. 12, 1693).

16. "Je ne voudrois jamais m'amuser à ces differentes natures de chaines, que Mr. Jo. Bernouilly propose comme devant achever ou pousser plus avant cette speculation. Il y a de certaines lignes courbes que la nature presente souvent à nostre vue, et qu'elle decrit pour ainsi dire elle mesme, lesquelles j'estime dignes de consideration, et qui d'ordinaire renferment plusieurs proprietez remarquables, comme l'on voit au Cercle, aux Sections coniques, à la Cycloide, aux premieres Paraboloides et à cette *Catenaria.* Mais d'en forger de nouvelles, seulement pour y exercer sa geometrie, sans y prevoir d'autre utilitè, il me semble que c'est *difficiles agitare nugas,* et j'ay la mesme opinion de tous les problemes touchant les nombres. *Calculis ludimus, in supervacuis subtilitas teritur,* dit quelque part Seneque en parlant de certaines disputes frivoles des philosophes Grecs" (*OH* 10:132–3, Sept. 1, 1691).

17. "A l'egard des lignes de Mr. Bernoulli, vous avés raison, Monsieur, de ne pas approuver qu'on s'amuse à rechercher des lignes forgées à plaisir. J'y adjoute pourtant une limitation: si ce n'est que cela puisse servir à perfectionner l'art

d'inventer. C'est pourquoy je ne desapprouve pas que des personnes qui ont du loisir et de l'inclination, et surtout des jeunes gens, s'y exercent. Et c'est pour cela que je ne veux pas décourager non plus ceux qui s'exercent dans les nombres" (*OH* 10:160–1, Sept. 21, 1691).

18. "Vostre meditation pour les Tangentes par les foyers me paroit bien profonde. Elle suppose pourtant des choses qui ne peuvent estre admises comme evidentes. Et quoyque des tels raisonnemens puissent quelque fois servir à inventer, l'on a besoin ensuite d'autres moiens pour des demonstrations plus certaines" (*OH* 9:538, Nov. 18, 1690).

19. "Car ce que j'aime le plus dans ce calcul, c'est qu'il nous donne le même avantage sur les anciens dans la Geometrie d'Archimede, que Viete et des Cartes nous ont donné dans la Geometrie d'Euclide ou d'Apollonius; en nous dispensant de travailler avec l'imagination" (*OH* 10:227, Jan. 8, 1692).

20. *OH* 10:496ff., passim.

21. Huygens makes this painful admission to Leibniz (*OH* 10:129, 139, Sept. 1691).

22. When told of the discovery by l'Hospital, Huygens was still insisting that his method was equivalent to Leibniz's, but when showed the proof he confessed that he did not see how l'Hospital was led to the solution (*OH* 10:305, 307, 312–15, 325–35; July 26 to Oct. 22, 1692).

23. On Boyle and making grand syntheses, see *OH* 10:229, 239, 263. On Descartes, see his disparaging remark that Descartes, jealous of Galileo's fame, sought to make a new philosophy, *OH* 10:404.

24. *OH* 21:820–1.

25. For example, he remarks to l'Hospital that Leibniz has an immoderate desire to aggrandize his achievements (*OH* 10:439, Apr. 9, 1693).

26. *OH* 10:269, 476, 509.

27. For other examples, see Shapiro's discussion of Huygens's optics in "Kinematic Optics," including his conclusion: "The mathematics, though, is always given a physical interpretation, and it is this intimate relation between the mathematics and the physics that makes the *Traité* such an elegant masterpiece" (p. 231); and Costabel's discussion of Huygens's work on compounding motions in *Leibniz and Dynamics,* pp. 74–83.

28. Mersenne writes to Christiaan's father: "I can believe nothing other than that, if he continues, he will some day surpass Archimedes" ("Je ne croy pas s'il continue, qu'il ne surpasse quelque jour Archimede"; *OH* 1:47, Jan. 3, 1647). From that day forth, Constantijn referred to his son as "mon Archimede," even in letters to the French court.

29. Leibniz uses the phrase when he tells Johann Bernoulli that Huygens has died, and Bernoulli repeats it when he replies (*OH* 10:721, July 29, 1695).

Bibliography

Manuscripts and books printed in the seventeenth century are cited as completely as possible in the Notes where their reference occurs. Modern editions and secondary literature directly pertinent to the topics discussed, including the better biographies of Huygens, are listed here.

Agostini, Amedeo. "Notizie sul ricupero dei libri V, VI, VII delle 'Coniche' di Apollonio." *Periodico di Matematiche* 11 (1931):293–300.

Aiton, Eric J. *The Vortex Theory of Planetary Motions.* New York: American Elsevier, 1972.

Apollonius. *Les coniques d'Apollonius de Perge.* Translated by Paul ver Eecke. Brugges: Desclée, de Brouwer, 1924.

　Treatise on Conic Sections. Edited and paraphrased by Thomas L. Heath. Cambridge University Press, 1896.

Archibald, R. C. "Curves, Special." *Encyclopedia Britannica,* 14th ed., 6 (1929): 887–99.

Archimedes. *Opera Omnia.* Edited by J. L. Heiberg; 2d ed. Leipzig, 1910–15.

Ariotti, P. E. "Aspects of the Conception and Development of the Pendulum in the 17th Century." *Archive for History of Exact Sciences* 8 (1972):329–410.

Baron, Margaret E. *The Origins of the Infinitesimal Calculus.* Elmsford, N.Y.: Pergamon, 1969.

Barrow, Isaac. *The Geometrical Lectures of Isaac Barrow.* Translated by J. M. Child. London: Open Court, 1916.

Belaval, Yvon. "Huygens et les milieux Parisiens." In *Huygens et la France,* pp. 49–56. Paris: Vrin, 1982.

Bell, A. E. *Christian Huygens and the Development of Science in the Seventeenth Century.* London: Arnold, 1947.

　"The 'Horologium Oscillatorium' of Christian Huygens." *Nature* 148 (1941): 245–8.

Bennett, J. A. "Hooke and Wren and the System of the World: Some Points towards an Historical Account." *British Journal for the History of Science* 8 (1975):32–61.

　The Mathematical Science of Christopher Wren. Cambridge University Press, 1983.

Bernoulli, Jakob, *Opera.* 2 vols. Geneva, 1744. Reprinted, Brussels: Culture et Civilisation, 1967.

Bernoulli, Johann. *Opera Omnia.* 4 vols. Lausanne, 1742. Reprinted, Hildesheim: Olms, 1968.

Birch, Thomas. *The History of the Royal Society of London for Improving of Natural Knowledge.* 4 vols. London, 1756-7. Reprinted, London: Johnson Reprint, 1968.

Bortolotti, Ettore. "A chi dobbiamo il ricupero dell'opera di Apollonio su le Coniche?" *Archeion* 11 (1929):395-6.

"Quando, come e da chi ci vennero ricuperati i sette libri delle 'Coniche di Apollonio.'" *Periodico di Matematiche* 4 (1924):118-30.

Bos, H. J. M. "Differentials, Higher-Order Differentials and the Derivative in the Leibnizian Calculus." *Archive for History of Exact Sciences* 14 (1974):1-90.

"Huygens, Christiaan." *Dictionary of Scientific Biography* 6:597-613.

"The Lemniscate of Bernoulli." In *For Dirk Struik,* edited by R. S. Cohen, J. J. Stachel, and M. W. Wartofsky, pp. 3-14. Dordrecht: Reidel, 1974.

Bos, H. J. M., Rudwick, M. J. S., Snelders, H. A. M., and Visser, R. P. W., eds., *Studies on Christiaan Huygens.* Invited Papers from the Symposium on the Life and Work of Christiaan Huygens, Amsterdam, August 22-5, 1979. Lisse: Swets & Zeitlinger, 1980.

Bosmans, Henri. "Galilée ou Huygens?" *Revue des Questions Scientifiques,* Ser. 3, 22 (1912):573-86.

Bosscha, J. "Christian Huygens." *Archives Neerlandaises des Sciences Exactes et Naturelles* 29 (1896):352-412.

Boyer, Carl B. "Early Rectifications of Curves." In *Mélanges Alexandre Koyré,* 2 vols. 1:30-9. Paris: Hermann, 1964.

History of Analytic Geometry. New York: Scripta Mathematica, 1956.

A History of Mathematics. New York: Wiley, 1968.

Brugmans, Henri L. *Le séjour de Christian Huygens a Paris et ses Relations avec les Milieux Scientifiques français suivi de son Journal de voyage a Paris et a Londres.* Paris: Librairie E. Droz, 1935.

Bruins, Evert M. "On Curves and Surfaces in the XVIIth-XIXth Century." *Physis* 12 (1970):221-36.

Burch, Christopher B. "Christiaan Huygens: The Development of a Scientific Research Program in the Foundations of Mechanics." Ph.D. dissertation, University of Pittsburgh, 1981.

Christensen, S. A. "The first determination of the length of a curve." *Bibliotheca Mathematica,* Ser. 2, 1 (1887):76-80.

Coolidge, Julian L. "The Unsatisfactory Story of Curvature." *American Mathematical Monthly* 59 (1952):375-9.

Costabel, Pierre. "Essai sur les secrets des *Traités de la Roulette.*" *Revue d'Histoire des Sciences* 15 (1962):321-50, 367-9.

"Isochronisme et accélération: 1638-1687." *Archives Internationales d'Histoire des Sciences* 28 (1978):3-20.

Leibniz and Dynamics: The Texts of 1692. Translated by R. E. W. Maddison. Paris: Hermann, 1973.

"La 'loi admirable' de Christian Huygens." *Revue d'Histoire des Sciences* 9 (1956):208-20.

Crommelin, C. A. "Christiaan Huygens." *Christiaan Huygens* 17 (1938-9):247-70.

"The Clocks of Christiaan Huygens." *Endeavor* 9 (1950):64-9.

"De isochrone conische Slinger van Christiaan Huygens." *Physica* 11 (1931): 359-64.

"Sur l'attitude de Huygens envers le calcul infinitesimal et sur deux courbes intéressantes du meme savant." *Simon Stevin* 31 (1956):5–18.

Defossez, Leopold. *Les savants du XVIIe siècle et la mesure du temps.* Lausanne: Edition du journal suisse d'horlogerie et de bijouterie, 1946.

Descartes, René. *Oeuvres de Descartes.* Edited by Charles Adams and Paul Tannery; 13 vols. Paris: Leopold Cerf, 1897–1913.

Dijksterjhuis Eduard J. "Christiaan Huygens. An Address Delivered at the Annual Meeting of the Holland Society of Sciences at Haarlem, May 13th, 1950, on the occasion of the Completion of Huygens's Collected Works." *Centaurus* 2 (1953):265–82.

The Mechanization of the World Picture. Translated by C. Dikshoorn. New York: Oxford University Press, 1961.

"De Ontdekking van het Tautochronisme der cycloidale Valbeweging. Een Bijdrage tot de 300e Herdenking van den Geboortedag van Christiaan Huygens op 14 April 1929." *Euclides* 5 (1928–9):193–208.

"Over een mechanisch Axioma in het Werk van Christiaan Huygens." *Christiaan Huygens* 7 (1929):161–80.

Val en Worp: Een bijdrage tot de geschiedenis der mechanica van Aristoteles tot Newton. Groningen: Noordhoff, 1924.

Dobson, R. D. "Galileo Galilei and Christiaan Huygens." *Antiquarian Horology* 15 (1985):261–70.

Drake, Stillman. "Galileo and the Projection Argument." *Annals of Science* 43 (1986):77–9.

Dugas, René. *A History of Mechanics.* Translated by J. R. Maddox. Neuchatel: Editions du Griffon, 1955.

Dutka, Jacques, "Wallis's Product, Brounker's Continued Fraction, and Leibniz's Series." *Archive for History of Exact Sciences* 26 (1982):115–26.

Edwardes, Ernest L. *The Story of the Pendulum Clock.* Altrincham: Sherratt, 1977. (Includes an English translation of the 1658 *Horologium*.)

Edwardes, Ernest L., and Dobson, R. D. The Fromanteels and the Pendulum Clock." *Antiquarian Horology* 14 (1983)250–65.

Elzinga, Aant. *Notes on the Life and Works of Christiaan Huygens (1629–1695).* University of Göteborg, Department of Theory of Science, Report No. 88. Göteborg: Avdelningen for Vetenskapsteori, 1976.

On a Research Program in Early Modern Physics. Göteborg: Akademiförlaget, 1972.

Favaro, Antonio. "Galileo Galilei e Christiano Huygens. Nuovi documenti sull'applicazione del pendolo all'orologio." *Rivista di Fisica, Matematica, e Scienze Naturale* 13 (1912):3–20.

Fermat, Pierre de. *Oeuvres de Fermat.* Edited by Paul Tannery and Charles Henry, 4 vols. Paris: Gauthier-Villars, 1891–1912.

Foà, Alberto. "La Cicloide." *Periodico di Matematiche,* Ser. 4, 17 (1937):65–95.

Frankfourt, U., and Frenk, A. *Christiaan Huygens.* Translated (Fr.) by I. Sokolov. Moscow: Editions Mir, 1976.

Gabbey, Alan. "Huygens and Mechanics." In *Studies on Christiaan Huygens,* edited by H. J. M. Bos, M. J. S. Rudwick, H. A. M. Snelders, and R. P. W. Visser, pp. 166–99. Lisse: Swets & Zeitlinger, 1980.

Galilei, Galileo. *Dialogue Concerning the Two Chief World Systems – Ptolemaic & Copernican.* Translated by Stillman Drake. Berkeley and Los Angeles: University of California Press, 1962.

Two New Sciences, Including Centers of Gravity & Force of Percussion. Translated by Stillman Drake. Madison: University of Wisconsin Press, 1974.

Grootendorst, A. W., and van Mannen, J. A. "Van Heuraet's Letter (1659) on the Rectification of Curves: Text, Translation (English, Dutch), Commentary." *Nieuw Archief voor Wiskunde,* Ser. 3, 30 (1982):95–113.

Gunther, R. T., ed. *Early Science in Oxford.* Vol. 6, *The Life and Work of Robert Hooke (Part 1).* Oxford, 1930.

Haas, August. *Versuch einer Darstellung der Geschichte des Krümmungsmasses.* Inaugural-Dissertation zur Erlangung der Doctorwürde in den Naturwissenschaften. Tubingen: Fues, 1881.

Hall, A. Rupert. *From Galileo to Newton: 1630–1720.* New York: Harper & Row, 1963.

"Horology and Criticism: Robert Hooke." In *Science and History: Studies in Honor of Edward Rosen,* edited by Erna Hilfstein, Pawel Czartoryski, and Frank D. Grande, pp. 261–81. Studia Copernicana 16. Wroclaw: Polska Akademia Nauk, 1978.

"Mechanics and the Royal Society, 1668–70." *British Journal for the History of Science* 3 (1966):24–38.

"Newton on the Calculation of Central Forces." *Annals of Science* 13 (1957): 62–71.

"Robert Hooke and Horology." *Notes and Records of the Royal Society of London* 8 (1951):167–77.

Hall, Bert S. "The New Leonardo." *Isis* 67 (1976):463–75.

"The Scholastic Pendulum." *Annals of Science* 35 (1978):441–62.

Harting, Pieter. *Christiaan Huygens in zijn leven en werken geschetst.* Groningen: Gebroeders Hoitsema, 1868.

Heath, Thomas. *A History of Greek Mathematics.* 2 vols. Oxford, 1921. Reprinted, New York: Dover, 1981.

Herivel, John. *The Background to Newton's* Principia: *A Study of Newton's Dynamical Researches in the Years 1664–84.* New York: Oxford University Press, 1965.

Hill, David K. "The Projection Argument in Galileo and Copernicus: Rhetorical Strategy in the Defence of the New System." *Annals of Science* 41 (1984): 109–33.

Hobbes, Thomas. *The English works of Thomas Hobbes of Malmesbury.* Edited by William Molesworth, 11 vols. London, 1839–45.

Hofmann, Joseph E. "Erste Quadratur der Kissoide." *Deutsche Mathematik* 5 (1940):571–84.

"Johann Bernoullis Kreisrektifikation durch Evolventenbildung." *Centaurus* 29 (1986):89–99.

Leibniz in Paris: 1672–1676. His Growth to Mathematical Maturity. Cambridge University Press, 1974.

"Das Opus Geometricum des Gregorius a S. Vincentio und seine Einwirkung auf Leibniz." *Abhandlungen der Preussischen Akademie der Wissenschaften, Mathematisch-naturwissenschaftliche Klasse (1941),* No. 13. Berlin, 1942.

"Ueber die ersten logarithmischen Rektifikationen. Eine historisch-kritische Studie in vergleichender Darstellung." *Deutsche Mathematik* 6 (1941):283–303.

"Ueber die Kreismessung von Chr. Huygens, ihre Vorgeschichte, ihren Inhalt, ihre Bedeutung und ihr Nachwirken. *Archive for History of Exact Sciences* 3 (1966):102–36.

"Ueber Gregorys systematische Näherungen für den Sektor eines Mittelpunkt-kegelschnittes." *Centaurus* 1 (1950):24–37.

"Vom offentlichen Bekanntwerden der Leibnizschen Infinitesimalmathematik." *Sitzungsbereichte der Osterreichischen Akademie der Wissenschaften, Mathematisch-naturwissenschaftliche Klasse,* Ser. 2, 175 (1966):209–54.

Hooke, Robert. *The Diary of Robert Hooke (MA, MD, FRS): 1672–1680.* Edited by Henry W. Robinson and Walter Adams. London: Wykeham, 1968.

Huygens, Christiaan. *Oeuvres complètes de Christiaan Huygens.* Publiées par la société hollandaise des sciences, 22 vols. The Hague: Martinus Nijhoff, 1888–1950.

The Pendulum Clock or Geometrical Demonstrations Concerning the Motion of Pendula as Applied to Clocks. Translated by Richard J. Blackwell, introduction by H. J. M. Bos. Ames: Iowa State University Press, 1986.

Huygens et la France. Table ronde du Centre National de la Recherche Scientifique, Paris 27–9, March 1979. Introduction by René Taton. Paris: Vrin, 1982.

James, Robert C., and Beckenbach, Edwin F. *James and James Mathematics Dictionary.* 3rd ed. New York: Van Nostrand, 1968.

Korteweg, J. "La solution de Christiaan Huygens du probléme de la chainette." *Bibliotheca Mathematica,* Ser. 3, 1 (1900):97–108.

Koyré, Alexandre. "An Experiment in Measurement." *Proceedings of the American Philosophical Society* 7 (1953):222–37.

Landes, David S. *Revolution in Time: Clocks and the Making of the Modern World.* Cambridge: Belknap, 1983.

Le Noir (*sic*), Timothy W. "The Social and Intellectual Roots of Discovery in Seventeenth Century Mathematics." Ph.D. dissertation, Indiana University, 1974.

Leibniz, Gottfried W. *Mathematische Schriften.* Edited by C. I. Gerhardt, 7 vols. Berlin, 1849–63. Reprinted, Hildesheim: Olms, 1971.

Sämtliche Schriften und Briefe. Series 3: *Mathematisch-naturwissenschaftlich-technischer Briefwechsel;* Vol. 1. Darmstadt: Akademie der Wissenschaften der Deutsche Demokratische Republik, 1976.

MacLachlan, James. "Mersenne's Solution for Galileo's Problem of the Rotating Earth." *Historia Mathematica* 4 (1977):173–82.

Morpurgo, Enrico. *L'Orologio e il pendolo.* Rome: Edizioni la Clessidra, 1957.

Naux, Charles. "L'Opus geometricum de Grégoire de Saint-Vincent." *Revue d'Histoire des Sciences* 15 (1962):93–104.

Newton, Isaac. *The Correspondence of Isaac Newton.* Vol. 3, edited by H. W. Turnbull. Cambridge University Press, 1961.

The Mathematical Papers of Isaac Newton. Edited by Derek T. Whiteside, 8 vols. Cambridge University Press, 1967–81.

Pappus. *Collectionis quae supersunt.* Edited by Frederick Hultsch, 3 vols. Berlin, 1875–8.

Pappus d'Alexandrie: La collection mathématique. Translated by Paul ver Eecke, 2 vols. Brugge: Desclée, de Brouwer, 1933.

Pascal, Blaise. *Oeuvres complètes de Blaise Pascal.* Edited by Léon Brunschvicg, Pierre Boutroux, and Félix Gazier, 14 vols. Paris: Hachette, 1904-14.

Patterson, Louise Diehl. "Hooke's Gravitation Theory and Its Influence on Newton." *Isis* 40 (1949):327-41; 41 (1950):32-45.

"Pendulums of Wren and Hooke." *Osiris* 10 (1952):277-321.

Plomp, R. "The Dutch Origin of the French Pendulum Clock." *Antiquarian Horology* 8 (1972):24-41.

Pugliese, Patri Jones. "The Scientific Achievement of Robert Hooke: Method and Mechanics." Ph.D. dissertation, Harvard University, 1982.

Reverchon, Léopold. "Huyghens horloger." *Revue Generale des Sciences Pures et Appliquées* 27 (1916):105-12.

Robertson, John Drummond. *The Evolution of Clockwork.* London: Cassell, 1931.

Scheler, Lucien. "Les *Lettres de Dettonville* offertes à Jean-Baptiste Colbert." *Revue d'Histoire des Sciences* 15 (1962):351-65.

Scott, Joseph F. *The Mathematical Work of John Wallis, D.D., F.R.S. (1616-1703).* London: Taylor & Francis, 1938.

Scriba, Christoph J. "Gregory's Converging Double Sequence: A New Look at the Controversy between Huygens and Gregory over the 'Analytical' Quadrature of the Circle." *Historia Mathematica* 10 (1983):274-85.

James Gregorys frühe Schriften zur Infinitesimalrechnung. Giessen: Mathematischen Seminars, 1957.

Shapiro, Alan E. "Huygens' Kinematic Theory of Light." In *Studies on Christiaan Huygens,* edited by H. J. M. Bos, M. J. S. Rudwick, H. A. M. Snelders, and R. P. W. Visser, pp. 200-20. Lisse: Swets & Zeitlinger, 1980.

"Kinematic Optics: A Study of the Wave Theory of Light in the Seventeenth Century." *Archive for History of Exact Sciences* 11 (1973):134-266.

Sloth, Flemming. "Chr. Huygens' Rectification of the Cycloid." *Centaurus* 13 (1969):278-84.

Stevin, Simon. *The Principal Works of Simon Stevin.* Vol. 1, *General Introduction: Mechanics,* edited by E. J. Dijksterhuis. Amsterdam: Swets & Zeitlinger, 1955.

Struik, Dirk J. "Heuraet, Hendrik van." *Dictionary of Scientific Biography* 6:359.

"Outline of a History of Differential Geometry." *Isis* 19 (1933):92-120.

van Geer, P. "Hugeniana Geometrica." *Nieuw Archief voor Wiskunde,* Ser. 2, 7 (1907):215-26, 438-54; 8 (1908):34-63, 145-68, 289-314, 444-64; 9 (1909-11): 6-38, 202-30, 338-58; 10 (1912):39-60, 178-98, 370-95.

van Helden, Albert. "The Accademia del Cimento and Saturn's Ring." *Physis* 15 (1973):237-59.

"'Annulo Cingitur': The Solution of the Problem of Saturn." *Journal for the History of Astronomy* 5 (1974):155-74.

"Eustachio Divini versus Christiaan Huygens: A Reappraisal." *Physis* 12 (1970): 36-50.

"Huygens and the Astronomers." In *Studies on Christiaan Huygens,* edited by H. J. M. Bos, M. J. S. Rudwick, H. A. M. Snelders, and R. P. W. Visser, pp. 147-65. Lisse: Swets & Zeitlinger, 1980.

"Saturn and His Anses." *Journal for the History of Astronomy* 5 (1974):105–21.

van Mannen, Jan A. "Hendrick van Heuraet (1634–1660?): His Life and Mathematical Work." *Centaurus* 27 (1984):218–79.

Westfall, Richard S. *Force in Newton's Physics: The Science of Dynamics in the Seventeenth Century*. New York: American Elsevier, 1971.

Never at Rest: A Biography of Isaac Newton. Cambridge University Press, 1980.

"Newton and the Acceleration of Gravity." *Archive for History of Exact Sciences* 35 (1986):255–72.

Yokoyama, Masshiko. "Huygens and the Times-Squared Law of Free Fall." *Proceedings of the XIV International Congress on the History of Sciences, 1974* 2 (1975):349–52.

Zeuthen, H. G. *Die Lehre von den Kegelschnitten im Altertum*. Reprinted from the 1886 (Copenhagen) German translation. Hildesheim: Olms, 1966.

Ziggelaar, August. "Aux origines de la Théorie des Vibrations Harmoniques: Le Père Ignace Gaston Pardies (1636–1673)." *Centaurus* 11 (1965):145–51.

"Les Premières Démonstrations du Tautochronisme de la Cycloïde, et une conséquence pour la théorie de la vibration harmonique." *Centaurus* 12 (1967): 21–37.

Index